MARS 20: Saving Utopia

Shelby Hiatt

Mars 20: Saving Utopia
Copyright © 2016 Shelby Hiatt
All rights reserved.

ISBN-13: 978-1539634911

Official Report: CALAMITY 2055

1

We were dust surfing out on the dunes doing serious Alpine moves, carving competition level linked turns, bodies close to the ground, arm out, touching dust, raising spray, not a worry in the world. We didn't know it had started. I'm eighteen and my boyfriend Rob just turn nineteen and we never noticed the calamity before that dust surfing day.

We were testing the new protection gear for Lewis, our whiz kid genius teacher, shooting up in the air, looping in long, slow arcs and landing gently, not hard like on Earth--we have less than half your gravity. We were doing moves we could never do before because of the bulky gear we had to wear and weren't wearing that day which I'd better explain since this report is official.

Our Martian air is mostly carbon dioxide, not enough oxygen

to breathe and the temperature is always sub-zero, so what we need when we leave an airlock is warmth, protection from solar radiation, and oxygen; three things, that's all. The settlers, our parents, came here with those big, bulky space suits and bubble helmets, only a little better than your first astronauts. You've seen those pictures and watched them blast off.

So when Lewis showed us the new gear in the lab that day and said, "It's like skin, call it skin," we had high hopes for lightweight space suits and Lewis did not disappoint.

He's amazing.

"Is it warm?" I said to him.

"That's what you're going to test, but yeah. It's plenty warm."

Rob and I were holding our new silvery skins, so light and flexible it seemed impossible they could protect us, but Lewis knows what he's doing.

"By the way," he said. "There are cameras in the face glass and in the arms."

"That's so cool."

"It'll record everything you see and do."

"Like astronauts?" Rob said.

"Better."

We pulled on the skins, ready to go.

"The gloves are so thin I can feel through them," I said. "This is great."

Rob said, "The whole suit's great."

"Go test it."

So that's what we were doing.

Did I mention that Lewis had transformed the clumsy bubble helmet into a kind of pop-up hoodie? You grab it at the back of your neck, pull it up over your head where it pops into shape with oxygen channels, and the glass-like face cover clicks into lock position under your chin.

Best. Helmet. Ever.

You can turn your head and look around inside it. You don't have to turn your whole body. It's almost like not wearing any body gear at all with the suit so flexible. Parents and kids were going to love it.

So we rovered to the dunes, climbed to the top and threw down our boards. And that's where we were freestyling and testing and still didn't know about the calamity. We were launching off the dunes higher and higher and floating down easy to soft landings. We even trekked up to still higher dunes and did those Alpine moves in formation just for fun and that's not easy, even in skin gear and glass. After a while, we moved down the slope toward level ground doing lazy intricate moves, normal to us but they'd look slow and gentle to anybody on Earth.

We were both pretty spectacular that day with no idea the calamity had started.

Walking back to the rover Rob said, "Dad'll love this, he hates the heavy gear."

"Everybody does."

We went straight to the lab where Lewis was waiting for us. When we came in he didn't turn around eager to know how the test went like you'd expect. That's not how Lewis is. We're used to his quirky ways and we threw back the cool new hoodie-headgear (you can't imagine what an improvement that is!) and Rob walked over behind him and grabbed him in a bear hug.

"Fantastic. The best gear ever...*ever.*"

Lewis turned around, no smile, all business, and said, "Tell me."

"What do you want to know? Completely free movement. Audio clear, perfect at any distance."

"It works," I said. "And we gave it a tough test. Nobody but a surfer would put it through what we did." I held my arms out. "Look, no scrapes, no damage, and we were twisting and turning."

Lewis looked me over and then Rob, examining the surface of the gear.

"Comfortable?" he said.

"Very."

"Flexible?"

"Of course."

"Free movement?"

"Totally."

"Okay, strip."

We peeled the close fitting gear off our thermal body cover

which is the indoor wear for all of us, the same for everybody, completely comfortable. (We don't understand all the different clothes worn on Earth. Weather, I'm thinking?)

"This is huge," Rob said. "I hope you know that."

Lewis didn't seem to hear. He was studying the gear, going over it centimeter by centimeter, examining where the dust stuck and where it didn't; mostly it didn't. We'd lost him, which was all right because we'd done a big thing for him testing it.

Then just as we were leaving, wearing our old, heavier suits, Lewis said, "There was a call for you, Rob."

"Who was it?"

"Your dad. Said your audio is off."

Rob and I looked at each other. Bad sign. We didn't want our parents looking for us, worried about another test we might be doing for Lewis, and audio turned off was a dead giveaway. We walked out and left Lewis still quietly focused on the new thin-gear.

2

Now since this is an official report and everything has to be chronological, you need to know what was happening at this same time. The Twentieth-Earth-Year-On-Mars party for the original settlers was going on in the council dome. The big room was decorated, I'd helped with that the day before. Tables were loaded with particularly Earth-like food. A whole turkey formed of vegetables looked like the pictures I'd seen in memory pieces, golden brown chest with legs attached, strange looking to us kids--no animals on our planet. But the parents liked it and were eating it. Then an argument broke out.

Two parents, a couple, were arguing loud about something personal. That didn't happen here. No fighting. No arguments that kids ever heard.

"Well fuck you then," said a dad.

"That'd be a welcome change," said his wife.

Several women were trying to lead her away and she yelled back, "I'm not taking it anymore."

The dad followed her and said, "You won't have to, you bitch. I'm gonna..."

"Hey, hey, hey..." Other dads grabbed him.

Watching the tape the next day Rob and I got their drift but we'd never heard words like that spoken aloud. Neither had any other Gen2s, I'm pretty sure. Parents always kept their disagreements private.

But at the party with no Gen2s around, the dad lost it and yelled again, "That little shit, she's..." Somebody's hand went over his mouth.

Other arguments immediately flared up, hard to hear what they were saying the party was so out of control by then. Dr. Sylva, our beautiful psychotherapist, finally got everybody singing Happy Anniversary To Us--don't know how she managed to do that--and it went quiet again.

So all this was going on as we were peacefully surfing in the wonderful new skins and was probably still going on when Rob and I reached his hab's airlock door.

I said to him, "Want me to come in?"

He said, "Yeah. I might need some backup."

***

Rob's mother was curled in his dad's arms shaking and whimpering. Rob went over and hugged her.

"I'm here, Mom. I'm okay. Kit's here too."

She looked up at me kind of distracted, then put her head down sniffling. Rob's dad rocked her and patted her like he would a child.

"What happened?" Rob said.

"She's feeling low."

They weren't worried about us testing something for Lewis. This was something else. Bigger and worse. And his Dad holding his mother that way made them both look helpless.

Rob mouthed the words "what's wrong?" but his dad still wouldn't answer. Just shook his head.

Rob didn't know what to do. We rarely saw emotion like that. Settlers were very careful about not showing anger or sadness even though we could tell it was there and they were controlling it. That's what Rob's dad was doing, being strong and comforting his mom.

I felt like I should do something so I said to her, "Everything's okay. We're making things better. Lewis is, isn't he, Rob?"

"Yeah. New gear so thin he calls it skin, Mom. Not much heavier than our thermals. And the helmet's like a hoodie. You can go outside and walk around and do everything like you're indoors. You'll like it."

His mother stopped whimpering and sat up. She looked at us, red-eyed and miserable and we waited. I think I smiled at her, or maybe I just tried.

In a weak voice, she said, "I won't like it."

We didn't know what to say.

"I want to walk out the door and feel the sun," she said. "I want to smell green grass and have an outdoor barbecue without wearing body-gear at all..." She was tearing again and trying not to. "...and I want a flower pot outside a kitchen window that I can open..." She was gulping and choking. "...and I want a dog..."

She broke down and sobbed.

Rob's dad rocked her for a minute then he helped her stand. We could see this was embarrassing for her. She was a very dignified woman. She was also tough enough and in love with Rob's dad enough to come to Mars, a one-way, goodbye-Earth trip. Very brave of her. And yet she was the first to break. Well, the first we knew about, not having watched the anniversary party video yet.

We kind of backed away from her and she said, "I'm not crazy. You don't have to be afraid of me. I'm just a little homesick."

She managed a choking smile and looked better.

Rob said, "You know what, Mom? We'll get you that gear from Lewis right away."

But this time when she tried to smile she couldn't. You could see the thought of putting on gear, even something so light it's called skin, disturbed her.

There was an uncomfortable silence and I said, "We'll go help Lewis now, okay?" Rob and I both wanted out of there.

She nodded and was about to cry again so we went to the door.

I heard Rob's dad quietly saying to her, "You'll be okay. Just gotta buck up..."

From the sound of his voice, this wasn't new, he'd been aware of this kind of thing for a long time. But it was new to Rob and me, and for us, it marked the day the calamity started. We thought it was homesickness.

But then, a week later we discovered how bad it really was because you don't call a brawl at a baseball field homesickness.

3

Bats and balls and gloves, everybody in the new skins and glass, parents and players loving it. The players walked around before the game swinging their arms, pitching back and forth, running in place which is more like hopping in place in our 0.4-gravity. A couple of them did somersaults ending with their arms out like an Olympic finish, the best part of any Olympic memory piece. They got applause from the bleachers.

Somersaults were impossible in our bulky outdoor gear a week earlier, so the applause was really for the skin and for Lewis who wasn't there--he's not interested in baseball. Somebody turned a cartwheel, got applause, and turned a couple more.

Things were looking up. They were happy parents that day on our baseball field which, by the way, is huge.

"Has to be big to make home runs reasonably difficult," they explained.

On Mars, a ball flies a long time before it lands, like golf on your moon. You know about that.

They even put in AstroTurf or something like it that Lewis cooked up to make the outfield more authentic. Red Mars regolith was fine for the infield but they made the outfield green, grass-like.

They built bleachers that they said were like the small fields in hometowns all over American.

We told them, "You don't have to do that."

We'd never seen a hometown baseball field except on a wall screen. But then we realized the game and the field and the bleachers and AstroTurf were for them more than for us, and that was all right. That was years ago and this was only a week after the 20th-anniversary party and our dust surfing in the new skin and glass.

"Okay, kids. On the field!"

Before the adults played, there was always a game for the little ones. They were wearing skins too, moms and dads watching. Up in the bleachers, I saw Rob's mom looking normal, his dad with her being protective, not sitting with the ball-crazy dads down close. The permafrost made a real dugout too much work so they built a little hutch above ground and that's where the enthusiastic dads were.

"Okay, let's go, let's go," they were saying, energetic, pacing and clapping their hands.

Rob and I were out with the kids who were so young and small they needed a couple of teen players to relay the ball in from the outfield to keep the game going. These young players, eight Earth years old and up, love playing but they tended to drift around from team to team and that drove the dads crazy.

One dad called out, "Look alive, you guys."

Another one called out even louder, "Go get 'em Orbiters."

But Orbiters or Comets, these kids didn't care all that much about their team or the score. Even when their dads talked to them about fighting, team spirit the kids just couldn't get into it. Don't get me wrong. They wanted to win, they just didn't feel competitive. The playing was the fun, that's all, even for Rob and me.

The game started, going fine, having fun. A couple of kids collided out in the field running for a fly ball. They got up, "sorry," "sorry," and like always, back at the hutch/dugout, dads were shaking their heads, *hopeless kids*. They wanted to see spirited rivalry like on Earth, competition not "sorry." But we went on playing. Nothing unusual. Yet.

Then another solid hit came floating way out to left field and by the time Rob and I relayed it in, two players had scored. The thing is, the second run was pretty clearly out. We could see it out in the field. They could see it up in the bleachers. But the dad who was the umpire called his daughter, Bethann, safe.

I'm guessing you can imagine what happened.

Everything went quiet for a second because it was so obviously a bad call, then several dads leaped out of the hutch and all hell broke loose. Pent-up fury exploded.

Several dads ran out on the field, surrounded the ump and yelled at him and at each other with so much kinetic energy their feet came off the ground. They weren't asking him to talk it over and maybe rethink what was obviously a wrong call. No, no. They were yelling, gesturing, calling him and each other names. Parents in the bleachers stood and yelled too, even the moms, as loud as the dads.

Okay. The ump called it in favor of his daughter's team and

that's not good. He shouldn't do that. But it's a game, *a game.*

One of the dads threw a punch just like we'd seen on screens and that started it. Arms were flailing, more punches, face glass and skin gear protecting them which was frustrating and made them go at each other harder.

Several moms ran out on the field and joined in. They started throwing punches just like the dads. I couldn't believe it. None of us in the outfield could. We stood there and watched our grown parents fighting each other, yelling obscenities, calling each other names although sound doesn't carry very far on Mars and that softened the ugliness. But it was worse than the anniversary party by far, and we were all watching, teens and kids.

Bethann came running out to us in tears and so did the other kids on her team, crowding around Rob and me for comfort, watching the brawl.

Rob and I had no choice but to leave them and start pulling the parents apart, kids watching. We pushed and blocked some of the swinging arms, got hit ourselves, and still it went on. It only broke up when one of the flailing dads hit me in the face glass and realized who I was and what was happening.

"Oh, I'm..."

He never said sorry. He just came to his senses, stood there dazed, then started helping us, pushing people apart and shouting at them until everybody backed off, panting and gulping. They looked around and realized how offensive this was, how bad it

15

looked, and worse, how incomprehensible it was to their kids.

The little ones were watching from a distance, shocked. The eight-year-old twins, Gaby and Noelle were crying. Their mother hurried out to them and hugged them and led them away. Several other smaller kids were crying and their parents went to them and tried to explain. I think the dads were still unaware that some of them had smears of blood inside their face glass. Terrifying.

When Rob and I saw the fights that break out on the field during your speedy Earth ball games in memory pieces, we thought they were funny. We laughed. We didn't take them seriously. One time I said to Rob they must be on the team "because they're good at putting on a violent act."

We didn't know.

To us it was entertainment and we weren't going to have a pop quiz on Earth behavior, so we always let it go. But seeing it that day on our ball field...

Eventually, most of the parents dribbled off, back into the colony, leaving Rob and me and a small group of kids who didn't want to go home. I think they felt better on the field with their friends and us.

I noticed Rob's dad was helping his mother make her way down the empty bleacher steps, shaken, her face hidden against his chest. They were the last to leave.

Rob and I got the game going again. There's no doubt the brawl shocked them but we were able to get the remaining kids batting and running bases again and having a good time.

4

"Not brawl proof," Rob said next day to Lewis.

"No way to make face glass brawl proof," Lewis said. "Your head moves around in there. If you get slugged really hard..." He shrugged. "How did it start?"

We told him.

The baseball field was a big red flag though we still didn't realize what we had on our hands. The next and final red flag was the picnic.

All right, I know what you're thinking. How do you have a picnic on Martian red dust? I have to ask you something. How do you have a picnic on Earth's green grass? It's full of insects and you're sitting on unsanitary ground with no protection whatsoever, birds and bugs flying around in an atmosphere full of pathogens. That's how it is, isn't it? And that's risky if you don't mind me saying it. But I know you have it worked out so I wish you many happy picnics. Our picnic was very different.

First of all, at the parents' request, a picnic is what we call any trip away from the colony. So I told Mom that Rob and I were going to take Zelda and her little brother Ben on a picnic with us.

"To Beagle 2," I said. "Not very interesting but Ben hasn't seen it. He'll like it." My mother looked nervous and frowned. "Zelda and Ben, Mom."

She knows very well they're our long time friends.

"Yes but..."

"What?"

"You might..."

"What?"

We never got in trouble. There isn't any trouble to get into, nobody here but us. Mars is crime free. Nothing dangerous in the dark or the light. And she has Homing, an app on her device that lets her redirect my rover back to our hab. All parents have it. It overrides anything the driver might do. We all have our own rovers, had them since we were kids. Homing and 360 radar make them safe. It's very safe on Mars. What was bothering my mother?

"Mom?" I said.

No answer. She didn't seem to know what she was doing or thinking. She seemed dizzy. I made her sit.

"Are you nauseated?" I asked, and made her put her head on her knees. "Ben's almost eight Earth years, Mom. You know how clever they are at that age, smart as a whip..."

"All of you are."

"Right, so we're not going to have any trouble. We'll be in touch with you the whole time. You can talk to me anytime you want. We won't be far."

She nodded. She knew all that but she looked like she was going to cry. I brought her water and asked her if she'd like to watch something.

I switched on the wall screen and started the crawl: movies, sports, theater, dancing--memory pieces made especially for settlers to make them feel "comfortable." Those did more damage than good I'm sure. (Let this report be my opinion on that.)

As the titles moved up the screen she said, "They're so beautifully made."

Putting Up Christmas Trees...Kids Playing in Parks...Hometown Fourth of July Parades > select your town.

My mother is a trained biologist and expert botanist, an intelligent woman who still reads and studies regularly and yet the glimpse of a Fourth of July parade, which she never went to on Earth she told me, was irresistible.

"Let's watch it," she said.

Her life was in the grower dome, the biosphere, where her expertise is key to our survival. She knows everything about the closed carbon dioxide/oxygen loop between humans and plants in our hydroponic farm. Why would she be interested in a Fourth of July parade?

I didn't say any of that to her, only thought it as I selected the parade for her and slipped away.

I called Rob and spoke softly, "She's lying on the couch. Looks like she's about to drop off."

"What's she watching?"

"A marching band playing Sousa in a parade."

19

"Sounds a little like my mom."

"I thought that too. What do we do?"

"Go to Sylva," he said. "I've talked to her about my mom."

***

An hour later I was nowhere near Dr. Silva. Rob, Zelda, Ben and I were on our "picnic" at the U.K.'s Beagle 2 site. We got out and walked around it: four solar plates, two lying flat, and two that never deployed blocking the antenna. That's just bad luck. No wonder they couldn't make initial contact. It's mostly covered with dust now.

"I'm surprised it isn't totally covered," Rob said.

Ben wanted to step on it and touch it, smart, curious boy.

"Think we have touching rights?" Zelda said.

She's the brightest kid of us all, a prodigy. Lewis knows it. He's training her to take over for him.

"It's U.K. property," she said, and decided Ben should only step around it carefully which he did circling the unit making footprints in the dust.

He looked back and grinned.

It made me remember Mom when she was happy, talking about playing in the snow and making snow prints, angels she called them. I never understood how walking around in cold snow then lying down and waving her arms could be fun exactly.

It was the print it left, I know, like Ben and his footprints. It's just that Mom doesn't understand the fun of walking around in C02 atmosphere and making red dust footprints. Guess it's not the same, not angelic.

"Whatcha doing?" I said to Ben.

He was squatting, studying the panels like a scientist. He blew on them. The dust raised and hovered. He frowned.

"They didn't write anything on it," he said. "Hello or something."

"They knew there wasn't anybody here to read it, right, Rob?"

Rob was looking at somebody approaching, a rover coming at us full speed.

Ben didn't notice any of that and said, "Do you think something's written on the other side maybe?"

"Maybe," said Zelda and told him he'd better not move it. "It's U.K. property. Better get permission."

"Ronnie's dad is English. We could ask him if it's okay."

Rob, still watching the approaching rover, said, "Who knows we're here?"

The rover was speeding toward us near our dust trail still hanging in the air. A double trail is very pretty, layers of pink hovering before they settle. But that day we weren't enjoying the trails. We were wondered who was spewing out that rooster tail and coming directly at us so fast.

"Doesn't look like a rover," said Rob.

Zelda watched and said, "It's a mono. It's Mr. Elkins."

"You think?"

"Sure."

Looking closer we could see she was right. It was a mono, the only one on the planet, and Mr. Elkins has it. He's blind, went blind right after the first settlers arrived.

Everybody was having eye problems in space and NASA knew the settlers would too when they came here, so they were sent with a treatment. But the treatment didn't work very well and maybe because he was older, Mr. Elkins went blind before Lewis came up with the antidote. It was too late to help Mr. Elkins but it saved everybody else's sight. See why Lewis is our most important settler?

As for Mr. Elkins, he always says, "I'm not only the oldest on Mars, I'm the blindest."

He was coming straight at us in the mono rover that our engineers built for him; four wheels, two close together both front and back. In the dusty Martian sand, it handles like a motorcycle which is what Elkins rode on Earth and what he wanted here. Like all rovers, his mono has radar and GPS talking to him all the time so he knows where he is. He goes buzzing around everywhere happy as a dolphin popping up all over the place, never lost, no collisions. Lewis interviewed him regularly for NASA/JPL so they could see how being blind on Mars compared to being blind on "wild and crazy Earth," as Elkins calls it.

"I can do everything you do down there except see," he always said.

Note for Earth readers: Most settlers continue to refer to Earth as down, ignoring the fact that we're on a solar plane and *out* from Earth, not *down.*

At Beagle 2 that day, he came up to us and stopped. He didn't get out of his rover.

Rob greeted him, "Mr. Elkins."

Elkins said, "What are you doing way out here, Rob?"

"Looking at Beagle 2. Ben's never seen it."

"Okay, but you gotta get back. Your dad..."

"My dad? What's wrong?"

Rob's dad and Lewis were the best scientists we had, both of them indispensable and solid with no signs of breakdown. Let me remind you that Rob's dad was not in the baseball field fight. He wasn't involved in sports at all. He worked with the engineers and with Lewis and in my opinion was the best adjusted of any settler. So what Elkins was saying about Rob's dad wasn't making sense.

"There's a problem." That's all he'd say.

"Why didn't they just call?"

"No, no. They wanted me to...they want you to come back."

That sounded bad. Rob didn't say anything and neither did the rest of us.

Elkins said, "Come on."

Rob insisted. "Tell me what happened."

But Mr. Elkins was already turning his mono and we were meant to follow.

5

At Rob's hab his mother was lying on the bed unconscious.

"Mom...Mom..."

"Let her sleep," Chris Chang said. He's our primary care doctor. "Don't wake her. I gave her something."

"I thought it was about my dad. What happened?" No answer. "What happened to her?"

"It's not about her."

This was confusing and nobody was clearing it up. Rob looked around.

"Where's my Dad?"

Elkins made Rob sit. Zelda and Ben had gone on home and I stayed along with a few other people, neighbors, and monitors who take care of colony problems. Dr. Chang started talking quietly to Rob.

"We all know what a good man your dad is," he said. "And how he's been taking care of your mother lately..."

"...where is he?" Rob was angry by then and I didn't blame him.

"I'm getting to that," Chang said. I knew this had to be bad.

Rob said, "Please tell me."

"Your mother was having another spell, a really bad one,

24

running and bumping into things and saying she didn't want to live any longer. So your dad called me and I came right over."

"Where is he?"

"He was lying on the ground outside the airlock door when I got here, no protection, only his thermal. Your mother doesn't remember how it happened but I could see. There are foot prints. Your mother was struggling, already outside..." Rob got a terrible look on his face. "She told him she wanted to die and was trying to go out without protection and your dad was trying to stop her, she told me that much. She doesn't remember anything about the struggle at the door but it's obvious what happened. She got out, and he went out and shoved her back in. The door shut and wouldn't open again for him. His sleeve was stuck in it. I have the piece of thermal that tore loose. Maybe it tore loose when he fell."

Death by CO2. Lungs can't sense the lack of oxygen. You get sleepy, pass out and die. It's painless, almost instant.

Rob stood still then he turned away and went into the airlock. I followed. We didn't say a word. His face was shiny with tears. He pulled me close. We wrapped our arms around each other and cried, both of us.

His favorite person was gone, his buddy, his dad. My favorite person was pained to tears. After a while something in us was eased. I don't think we'd cried since childhood.

When we stopped, Rob cleared his throat and said, "We

should go back in."

His voice wavered but he was standing straight, and with his arm slung across my shoulders, that's what we did.

Chang was right inside the door.

"We'll be taking her to the infirmary," he said gently. "She needs to rest." He took a good look at Rob and said, "Are you okay?" Rob nodded and Chang said, "Come visit her tonight."

***

That night we went with Dr. Chang into the infirmary's back room where Rob's dad's body lay. We stood looking at him. Tears jumped to Rob's eyes then to mine. After a few minutes Rob moved the sheet up over his dad's face, kept his hands spread on the covered face--last contact--and lowered his head.

When he removed his hands and straightened we both wiped our eyes and went back in to his mother.

She looked a little better, her head tilted back on a pillow. She was pale but she smiled when she saw us. Her voice was weak.

"Hello," she said. Rob leaned in and kissed her. "How's Dad?"

Rob turned to the nurse. "Could we have chairs?"

As we sat, his mother repeated it. "How's your dad doing?"

"He looks like he's asleep."

"I hope he sleeps well," she said. "I've been trouble for him you know."

Rob didn't let his voice quaver. "But you're better now, aren't you?"

"Yes. I don't know what came over me."

"Just a bad time. Everybody has bad times."

She looked at me. "How are you, sweetie? What are you writing and drawing these days?"

"Just stories about us. It's fun."

She smiled at me and took Rob's hand.

"You're the best kids..."

Her voice caught and she fought back tears.

"It's okay, Mom, everything's okay."

She smiled and closed her eyes and Rob stroked her forehead.

"Sleep now."

She was asleep in minutes and we left. Outside the infirmary Rob leaned over, and hands on his knees, he moaned. Little, quick sobs came out of him and before I could touch him he breathed deep and stood up again.

"I'm okay."

He took more deep breaths and we started walking to his hab.

"I'm thinking...it might come back to her," he said. "And if she remembers I don't think she'll be able to take it."

"Maybe it won't happen if she stays in the infirmary."

"She thinks Dad's coming to visit this evening."

6

In class the next morning, the first death on our planet was announced and the little children seemed to understand it better than the older kids and adults.

The question was, where should a dead body go? Where should it be "laid to rest" as settlers say.

"JPL talked about this before we came here," Lewis said. "They had several alternatives. Burial's no good; permafrost. And we've got no cremation chamber. We could build a room for casket size vaults but not in time for Rob's dad."

"Wrapped up outside, that body'd be frozen solid," I said. "He could wait for a vault room."

Lewis frowned. "Settlers wouldn't go for him being frozen outside."

The youngest ones were wide-eyed and interested. Rob wasn't in class or expected to be. Several parents were with him to keep him company doing everything they could, making his favorite food and talking about what a great guy his dad was.

He texted me: I want out of here. See you after class?

Lewis saw me texting and said, "Is that Rob?"

I nodded, "He wants to get out of his hab."

Lewis understood. He was always on our side in something like this. He came here when he was only seventeen, made a false birth certificate to get on Mars One. He's brilliant and much more like us than like the settlers.

He understood why Rob wanted to be in class with us. It's like another home, fifteen of us, four to nineteen Earth years, a one-room school house. Lewis teaches the older ones math and science. The rest of us who aren't interested in the sciences are in the scribes and sketchers group which includes me, working on stories and illustrations, writing little dramas and acting them out. It's all based on right and left brain learning. The older ones teach the younger ones how to read and write and do basic math. It works so well most kids don't want to go home at the end of the day. All their friends are there and they're playing together, making things, doing experiments.

See why this was where Rob wanted to be, with us doing pleasant things, not mourning his dad?

"We're not avoiding the death," Lewis said to me. "Look at the little ones."

The youngest were happily drawing pictures of a body underground, hands clasped on his chest, eyes closed and smiling. Others drew pictures of Rob's dad in the air sailing out into infinite space to join the asteroid belt forever. Rainbow colors. Nice.

"Looks like Rob's dad's either going to be under Mars regolith or out in space," I said.

"We'll see," said Lewis.

Jeff, our main guy at JPL was pinging him.

JPL: Deeply sorry to hear about this death.   He's a big loss to us and to your group, we know that. His relatives here are notified and it has made the news so it's important to make it an Earth-like burial. Send us video. Death on Mars should not be more off-putting than here. Your summer permafrost isn't so deep. Make it like an Earth funeral.

MARS: Will do. That seems to be what the settlers want. But the Gen2s including his son are wide open to something more practical. Whose wishes do we follow?

JPL: Ours.

Lewis texted Rob who responded: Do what you have to do.

Rob was not unfeeling, Lewis knew that. He knew what Rob meant, that what remained of his father was inert matter, ready to go back to the stars where it came from.

But NASA/JPL ruled so it was accepted that there would be a funeral as much like one on Earth as possible. Rob's dad was a settler and would be buried like one. Lewis announced it just before the end of day. The very youngest, the kids who drew the body smiling under red Martian soil, grinned. Their idea had prevailed.

I asked Lewis about Rob's mother. "She won't be at the funeral, will she?"

"No, no. It's a few days off and she'll be kept away."

She went silent the evening Rob's dad didn't show up. She no longer spoke and was fed intravenously.

One of the youngest boys in class came over to me with color sticks and paper and said, "I want to make a picture of clouds for my mommy. She's sad. How do they look?"

I sketched a heart shaped cumulus cloud for him to color and said, "Make the sky blue for her, not red, okay? Blue sky. She'll like that. And leave the cloud white or maybe pink."

His eyes lit up. "Like a sunset?"

I nodded. He'd seen the memory pieces. He knew his mommy's favorite sky colors. He went off with his paper and color sticks.

I started to leave but Lewis stopped me and drew me aside.

"There's trouble you're not being told."

"What trouble?"

He looked around and leaned close.

"Parents. The fight on the baseball field and Rob's mother's episode..."

"What?

"There's a lot of that going around and Chang and Sylva are keeping it quiet. They don't want to upset the kids."

"Why are you telling me this?"

"To get it off my chest. I'm not the group shrink, that's Sylva. I've told Rob too. Go help draw clouds," he said. "And don't mention anything about this."

Do you see how long it took us to put all this together?

Then it came, the undeniable wake-up for all of us: Ravi's dad during our dune race.

7

Every year we have a big dune race for Gen2s, and since a year on Mars is almost twice as long as on Earth, it's a much-anticipated event. Everybody comes out to watch; personal rovers are parked all over the place. Our rovers have governors on them and every self-correcting safety feature known. So like Mr. Elkins' rover, they're bumper cars with invisible radar for bumpers. You cannot get hurt. Kids have rovers and race around in them every day.

So there are always a lot of rovers parked outside the race area and this recent race was no exception with some people up in jetpacks circling and watching.

"Like a Fourth of July parade," Mom said.

Lots of American parents agreed. Why did they miss that parade so much?

We were out there racing, all the Gen2s over ten, speeding around a dune course that Lewis and Milo had laid out. Milo is a settler and designs and maintains our rovers. For every race, he disables the speed governor and the safety features. They'd keep us too far apart. For this year's race, he made additional modifications. He replaced the wheels with small skis--skids he called them--and put a single propelling track wheel in the center.

That changed everything.

"Gives the advantage to dust surfers," Rob said when we were practice driving before the start.

But everybody had the same amount of practice time. It was fair. Definitely an advantage to dust surfers though.

One more time my mother said, "No pit stops for fuel like on Earth," and shook her head and smiled at her little joke.

My biologist mother knew very well the big nuclear reactor above us (the Sun) is our energy source, good for another billion years or so. And like everybody else, she saw the solar arrays scattered all over the colony. Still, she made her joke and I wondered why she kept remembering Earth things like pit stops for fuel after all this time.

The race started. A parent waved a flag and we took off.

8

All the rovers were engineered the same which made the competition between driver's skill, not the rovers. There were markers for the boundaries of the course but they were far apart leaving plenty of latitude for the fastest route.

We were turning, twisting, looping over dunes, skidding down, then out across the flats. Dunes are a particular challenge, how to take them; straight across and risk bogging down, or side-winding which is faster but requires more skill. Yes, dust surfers did have the advantage.

All of us soon found the best, most comfortable way to take

33

turns and get across open areas depending on our driving skills. The course was long, had to be because of our speed, and we'd be out of our parents' sight for several minutes. Then we'd come around again and fly by in a different order and they'd cheer, somebody new leading. At one point the leader was me.

I'd never won. I was using my surfing skills. Then excellent surfer Rob led a circuit. I regained the lead and it looked like I might actually keep it and win. I was being careful, skidding out less often than others and staying on the course, managing dunes carefully. It was mostly luck being in the lead, I knew that, but the next time we went past the parents, Derek was almost touching one of my skids trying to pass me.

Derek, a Gen2, is Lewis's right-hand helper, half his age but already as brainy. In the dune race he's an aggressive driver, always a contender, and on that lap he was right behind me, very close. Too close.

He saw he was too close and quick backed off making Ravi bump him from behind. Only a bump, but it pushed Derek forward again into me, pushed me in front of somebody else--I don't know who--and we went flying and tumbling, four rovers spinning and skidding slowly. It was a mess. But we were each strapped into our roll cage and came to bouncing, sliding, rolling stops, unharmed.

We looked around and climbed out. Parents came running.

They uprighted a couple of overturned rovers and crowded

close. There has never been a pile-up like that in ten years of racing so I guess it looked awful to them but it wasn't. (Skids for wheels are sure to be banned.)

The other racers had gone on around the circuit and somebody else would win, not me or Rob or Ravi. They were already out of sight. Parents were worried and hovering but nobody was injured, not even bruised. Not enough Mars-g for that.

Even so, a mother suddenly said, "Get out! Get away from the rovers," like something might explode.

But this wasn't Monaco with cars carrying gas tanks, virtual bombs that really can explode in an oxygen loaded atmosphere-- we know something about your formula one races, thanks to sporty dads and sporty memory pieces.

Parents kept asking, "Are you sure you're all right?" "No bones broken?"

Do you see, NASA/JPL? A perfect example. Earth fears and Earth consciousness weren't gone or even lessened. They were flaring up like I hadn't seen in a long time.

Parents hugged us, teary, some of them. And others wouldn't let their kids get back in their rovers and finish the race.

Rob and I just went back and watched mild and generally quiet Zelda win. She's a tiger in her rover. We whistled and cheered for her. After her victory lap, Ravi's dad came over and glared at those of us who'd been in the pile-up.

"Who started it?" he said. He's a very mild man but at that moment he was on fire. "Who's responsible?"

We didn't know what he meant exactly, it was so obviously a pile up, an accident. Anybody could see that.

"Maybe me," I said. "I skidded into Ravi when I got bumped..."

"...I bumped her," Derek said.

The others spoke up quickly and we figured out the sequence of the accident for him: Derek into me, me into Ravi, Ravi into a couple more ahead of him...it wasn't that complicated. But Manoj, Ravi's dad, was irate.

Shaking his head and breathing hard he said quietly, which was creepy, "Ravi would have won, you know. He would have pulled ahead and won."

I flicked my eyes over to Derek. He raised his head in response, *Yeah, this is strange*, but we didn't say anything.

Manoj kept breathing hard and I wondered if there was some malfunction in his new skin and glass, not enough oxygen or too much. He just kept nodding his head like he knew his son would have won.

Poor Ravi was embarrassed. His dad wasn't like that at all. I didn't know him well but we all have a pretty good idea what the other colony members are like and Manoj was not unbalanced, we'd never seen it, anyway. He was a serious, biosphere engineer working in the grower dome. That's what we knew about him until the dune race.

Then when I remembered what Lewis told Rob and me as a secret, I began to get the picture. This strange behavior from so many parents including the brawling on the field and the 20th-anniversary party was not rare. It was common and was kept secret and had been going on for a long time.

9

"Manoj had fire in his eyes," I told Lewis. "He was furious."

Lewis never saw races or had any interest in them apart from the engineering of the rovers and he frowned.

"It's true," I said.

"I believe you. It's coming out."

"Is this the secret you told Rob and me about?"

"Yeah."

He didn't go back to work. He just stood there, staring at a corner, disturbed.

I went to Dr. Sylva in the infirmary. A psychotherapist would know about these things. Surely she'd seen it and treated it.

"Oh, yes," she said. It's..." She stopped herself. "Have you seen much of it?"

"No. Lewis told me and Rob about it and we haven't told anybody else. But the race...you should have seen."

"I heard. There's plenty of that kind of thing going around, but it's unexplained so far. We don't want you kids to worry about it."

"But the kids saw it," I said. "Everybody at the wreck saw it. This big smash up, nobody hurt, and there was Ravi's dad furious, accusing us of...I don't know what exactly. And that's not how he is."

She was nodding. "Irrational, confrontational, out of control..."

"Right."

"Problem is, we can't treat what we can't diagnose and it's worse every day. There are more outbreaks and they keep telling me they feel unsettled."

"Yeah. My dad said something like that to me. He said he didn't feel like himself these days."

***

Okay, the calamity was recognized.

Now, the sign. Believe me, it's important. And for that you need context.

New settlers were coming and everybody was excited. We'd waited years. They can come only when our orbits are close. They'd come once before. We had high hopes they'd give our parents the lift they needed, maybe even cure their homesickness which is the word Dr. Sylva was using while knowing it was something much more severe.

"It will take more than fresh settlers to cure this," she said. "But they'll help. For a while."

Of course, they would.

We'd seen how our parents acted when a new batch arrives. They rush out to the lander and hug them as they get off. They take them aside talking like they're old buddies, pumping them for news about "sports, movies, politics, new styles...just tell us everything."

The whole time the new batch is reeling from months of often weightless travel and then suddenly dealing with Mars-g and the locals hugging them. The parents take them back to the habs in rovers.

You'll each have a rover," they tell them and then add, "Light traffic today." Martian humor.

We'd seen all that and it would be funny and touching to see it again. Don't get the idea we Gen2s weren't interested in the new settlers themselves. We were, just not in their news.

Now stay with me on this. I'm getting to the sign and what followed.

Rob and I went up in jetpaks scouting the best location and found one not too far from the colony. A wide, flat, area you pass when you go north.

"Wait'll you see it," we said.

Once everybody agreed it was perfect, we started hauling bags of silica to the location, plenty of silica on Mars. The silvery white would show up well against the rust red regolith and Lewis added something to the silica so it would solidify. Our sign wouldn't blow away. But it would be covered with dust like everything else and Lewis would solve that too with embedded blowers on timers, solar powered of course.

We went to work, a line of loaded rovers moving out to the location past a line of empty rovers coming back. Bags of silica were stacked throughout the area. We laid out the shape of the letters with twine on stakes like you do on Earth. The letters would look like some cosmic giant had printed them with a giant stamp in giant white Helvetica, still a favorite texting font. Lewis thought a familiar font would make the newcomers feel more at home.

As we were working hard, the whole sign team out there, a rover came speeding toward us.

"Please don't let it be Mr. Elkins again with bad news," I said to myself.

This was supposed to be a happy, surprise mission.

Everybody stopped working and watched the rover come close.

It was Lewis.

Out of the lab?

Must be important.

I got an uncomfortable feeling. He brought his rover to a sensible stop--Lewis never swerved with a dust spraying flourish like the rest of us--and he got out with a dead serious look, more serious than usual. He didn't say anything for a minute. He saw the others with drums ready to roll and took a deep breath.

"Maybe don't do this."

"Do what?" I said.

He shook his head and looked down.

Rob said, "Why not, Lewis? What's up?"

It took him a minute then he said it. "No more settlers coming. Operation canceled."

Not a word from anybody. It didn't make sense. We were supposed to get new people every eight or ten years. That's why we were here, to populate the planet. Still, Lewis didn't say anything more so I did.

"We're alternative living space for settlers. We're supposed to be uploading people forever. What's going on?"

Rob said, "Yeah, what is this?"

Lewis took a deep breath. "They've found better planets, easier to terraform, easier to live on, something like that."

Rob smiled. "You're kidding. We know about those."

"It's something really new. Recent. That's what they're focused on."

"I don't get it," Rob said. "They can't do that."

"Yeah, they can. The real reason is that it's too hard on settlers. They can't send more people into a situation where they break down. They won't do it."

Rob walked over close to him, "They know what's happening, how the parents are?"

"I had to tell them, it's my job. I didn't know it would lead to this."

I guess we looked stunned because Lewis said, "They won't abandon us."

"What do you mean?"

By now the others had walked in close.

"What's going on?" Ravi said.

"No use making the sign," I said but I couldn't bear to say the rest. Lewis did.

"All future incoming settlers are canceled. For good."

It was quiet then somebody said, "Why? Parents love them."

Somebody else said, "We all like them."

When Lewis didn't answer Zelda said, "Are we going to be okay?"

Lewis nodded. "We'll be fine. We'll get everything we need, regular supply probes. Parents will have to learn to do without new settlers." He let that sink in then he said, "This isn't much change for you, but for parents, it might be...you know."

We knew, all right.

I said, "Have you told them?"

"It'll be announced later today. On walls everywhere."

I don't know what overcame me but I spoke up, loud. No more secrecy. I'd had enough.

"My Mom was in bad shape this afternoon," I said. "Anybody else's parents like that, breaking down?"

It was quiet a minute then somebody said, "Both of mine. They're trying to hide it."

"Mine too," said Zelda.

"Who else?" I said.

A voice said, "My mother was crying yesterday and my dad was yelling."

Another voice said yeah, the same kind of thing was happening, parents arguing a lot, then everybody started talking about it. It was happening in most of the habs, not anomalies, regular behavior.

"Wait," said Lewis. "Who has not seen this kind of thing?"

A few scattered hands were raised. He looked at all the Gen2 faces that he'd known since they were born and that he'd seen in class every day and said, "We have to deal with this."

10

It was like a class out there that day except Lewis didn't have a wall screen behind him to do the math. Anyway, math couldn't solve this problem. He wasn't comfortable but he went over what it was like for parents to look forward to settlers coming and I had to make myself remember he was a settler too, so it had to be hard for him and that got me thinking.

Lewis wasn't a victim of this sickness. He knew how parents

felt but it wasn't affecting him at all. Partly because he was much more one of us kids, our "old reliable Lewis" (not as old as other settlers) in the lab all day, Derek at his side and my dad their backup in engineering. Why wasn't Lewis affected?

"Settlers have lost their only contact with home," he said. "We have to understand what that's like. They're not going to shake a hand that was recently driving a car or was shopping in a mall..."

There was some laughter at that because memory pieces with malls were the only ones we could relate to, structures that safely enclose people like our modules, many with glass domes. Why aren't there more of those on Earth?

Lewis wasn't laughing.

"I guess that sounds odd to you," he said. "But believe me, it's not to them. Tactile things trigger memories, stuff they're used to. New settlers are as close as your parents come to all that and now even that's over, a big loss. It's like...if you had to go to Earth..." A lot of groans when he said that. "...yeah, right...and you could never come back here. After a while you'd be hurting and cranky too."

Lewis stopped for a second and let that sink in.

"Just remember," he said. "They still miss it. They were used to Earth and Mars is still new to them, maybe always will be. To you, it's the only world you know. To me, it's the one I like best. But to them..." He shook his head. "What we've got to remember

44

is that they want Mars to be more like home and incoming settlers are the next best thing."

We got it.

"Promise you'll let your parents know you understand, then help them if you can. Give them the respect they deserve."

"I don't think that'll help my mom," I said. "I never saw her like this."

"I've seen it on Earth," Lewis said. "Depression, sadness. It's awful. Be nice to them." Nobody said anything. "Try not to let them be alone. Dad's are going to try to hold out longer and be strong for the family but they don't have to. Nobody's going to go hungry here or be killed or hurt so bad they die. That's something your dads have a hard time with. So be patient and talk to them. And listen to them."

There was silence again and then we started talking among ourselves.

Lewis stopped us. "Hey. We've got a job to do."

Nobody moved. The sign no longer had any meaning. But then I thought about it and realized Lewis was right. This news was...what do settlers call it?...an inconvenient truth. Time to carry on. That's what my dad would say. Mom too.

I stood and said, "Let's get to work. Let's make the sign anyway."

The kids had dismal looks on their faces and didn't move. Then Zelda slowly stood. So did Ravi and Rich and Derek then some others, and finally they all got up, not happy but ready to work.

***

Back out on the field, Lewis joined in.

"Who's on barrel drums?" he said.

Somebody answered, "All of us now that the letters are laid out."

"All right, then."

Two or three to a drum we started rolling them back and forth, up and down and curving around to make sharp, well defined edges, laying down a thick coating of white silica on red Martian dust, working a long time until it spelled in permanent, giant, totally useless but visible from space, white Helvetica letters:

WELCOME TO MARS

11

It was late afternoon when we dragged back into the colony, tired but proud of ourselves. The feeling was short-lived. I'd almost forgotten what was coming. Lewis hadn't. It was his responsibility to convey messages from NASA.

He gave the signal to somebody at council and within seconds the message flashed on screens outside and inside habs overriding anything people were watching, accompanied by loud audio.

VOLUNTEERS FROM EARTH HAS BEEN CANCELED
NO FUTURE ARRIVALS SCHEDULED.

Blunt. No explanation. And since audio was only used for warnings, it sounded dangerous, not just unfortunate. It kept repeating to make sure it was heard and that was even more provoking.

Whose decision was audio and repetition? Worst possible way to break bad news.

People came out of their habs and workplaces looking around in disbelief.

"What's this?" they said. "What does it mean? Did you hear anything about this before?"

Of course, nobody had.

In anger, the largest number poured out of the grower dome spouting profanities, not asking questions. They marched as a group to council, went through the airlock, and throwing back their hoodie-helmets, walked straight into Thomas's office.

"What the fuck is this?"

Thomas didn't know what to say or do. He muttered something about following procedure, conveying NASA/JPL messages. He was a settler but not a parent, never paired off like most others. He had no idea what his badly stated, badly timed and badly delivered (audio!?) message would rouse. He didn't know these parents would come at him like he was personally responsible for the content and meaning. Or that they would be furious at his weak explanation: "...a message from NASA..."

Absolutely not enough. They didn't try to be reasonable like they might have in the past. They grabbed him and shook him.

He said, "I didn't realize...I didn't know..."

"Are you kidding?" one said.

"That's fucked," said another. "It's your job to know."

A dad got in his face and said, "That's what council's for, you moron. What part of No More Settlers did you not realize and know?"

A woman slugged him.

Another woman slapped a worker sitting at a desk. I was there by then watching all this.

It exploded into a riot.

They assaulted the council workers, or tried to, with flailing, wild punches. The workers ran into the back rooms and locked the doors. That angered the parents even more. I cringed against a wall.

That loud No More Settlers announcement was the straw that broke the camel's back and despite no camels being here, we Gen s understand the analogy perfectly. The audio announcement was the straw. Never again were parents going to see fresh Earth faces with fresh Earth news, something they'd been looking forward to for years. *Years.*

***

"Thomas blew it," I heard my dad say when I checked in at our hab. "Fucking thoughtless and cruel." He rarely got angry but he was then, though he stayed in control. "NASA must have kept it from us as long as possible."

Mom said, "They left it up to him how to announce it, I guess."

Dad made a growling sound which I'd only heard in memory pieces, never from him.

So the news was delivered. Badly. Although for the purpose of making this report honest, I don't know how else Thomas could have done it except no audio for sure. Maybe an electronic message to each colony member, silent and personal. Something they wouldn't hear all at once blared out at them from every angle and every wall screen. The shock should have been spread out over time, quiet and individualized. That's how I'd have done it, definitely not sudden, blunt, and loud. But I guess there was no easy way to hear news that bad.

They were furious.

The rioting continued into late afternoon, I watched from the classroom.

"You shitheads! You're all shitheads!" somebody yelled.

"Fucking idiots," from somebody else.

"Assholes...assholes...assholes," a woman kept repeating, her fists clenched. I couldn't see her face. She was looking down and walking aimlessly.

Who does she mean the assholes are, I wondered. NASA? JPL? Thomas? Everybody who dreamed up Mars One and Mars Two and made them possible? I was naive to think she had anyone specific in mind.

"What kind of sons of bitches are doing this?" somebody else yelled at the council dome like it would answer.

They're the ones that got you up here, I thought, and hover over every message we've ever sent them, and study everything we do. That's correct, isn't it, NASA? You're watching us all the time? And lots of the time our days are your nights so it's not a fun assignment keeping an eye on us: what we need, how we're doing, what should be sent in the next supply probe. Months getting it here, better be sure we need it and you haven't left anything out.

Not a fun, often sleepless job and you've done it for years. We know that. But that afternoon...wow.

Gen2s had never seen sustained anger on a scale our parents were demonstrating and it shook us. The younger kids watched through the classroom window ports as it got wilder, parents throwing anything that would break and trying to burn the modules though they know very well fire doesn't burn in Mars' atmosphere. The more things that didn't burn or break (modules are tough, they don't fracture) the angrier they got. The sun was about to set.

The kids were badly shocked and disturbed, very small ones crying. Rob and I got them away from the ports and told them not to worry because "everything was all right after the fight at the baseball field wasn't it?" That seemed to work and they settled down and began drawing or played games. The older kids chose to watch, some of them tearful.

It just wouldn't stop: frustrated parents fighting, swinging at each other without doing any damage wearing skin and face-glass, like on the baseball field. They backed off, panting and more frustrated. When the sun did set and it went suddenly dark--our thin atmosphere doesn't reflect twilight--the lights came on automatically all over the colony including inside the face glass.

Two men went on flailing with no damaging results-- extremely frustrating for them. After a while they stopped and stood looking at each other, light flashing across each other as they wavered. They started laughing, hysterical laughter I would call it, out of control. Then they collapsed on their hands and knees gulping and sobbing.

The women were in the same state, weak and out of control, stumbling and falling. Finally, everybody was sitting or lying on the ground, parent bodies scattered like sticks over the whole common area.

We didn't dare come out of the classroom. Rob and I watched

the parents lying in the dark, looking up at the stars, brighter than you ever see them from Earth, no twinkling in our thin air. Just big, bright stars looking down at our poor fallen parents.

Little Phobos came up and started his fast journey across the sky like everything was normal. Nobody moved. All eyes were on his just perceptible progress. In four hours he would traverse and be gone. The parents watched his odd, oval shape trundling along, and after a while, they began to sit up and stand and eventually they went back to their habs.

Many of them chose to go to the infirmary where one after another, they broke down completely, crying, holding onto a bed rail so they could stamp their feet and let off anger without floating up. They hated floating up. They wanted Earth-g, Earth-everything again, and it was never going to happen.

12

That evening after the riot it was like dominoes. There was no more covering up how they felt. Parents were so disturbed they were disabled and couldn't, or said they wouldn't, work. Nearly all of them had checked in the infirmary in defiance, angry at Mars and being stuck here. They weren't going to be involved in running it any longer.

Crazy thinking of course but not crazy to them. They were fed up.

In the infirmary there weren't enough beds, a new crisis.

"No time to try to figure out what's wrong with them," Chang said. "Gotta care for them even if we can't cure them."

His voice jiggled as he hurried from one place to another, me following, helping him slide the beds end to end along the wall to serve as seats.

Also helping us, Dr. Sylva said, "At least they won't be alone."

"Right," Chang said. "They can't isolate and we can study them."

The staff hustled around helping to set the last few beds in the walkways. There weren't enough for all the parents to lie down, but sitting two or three on a bed, they had a place to talk and mend their relationships which is exactly what they did.

Packed in like refugees, they drank cartons of milk and munched cookies, the only finger food on hand, and began laughing off the riot, relieved and friends again. But the crowding with no place to sleep would soon be a problem. Like...that night.

"They've gotta stay her," Chang said, hands on his hips, shaking his head. "We have to have beds."

Rob and I went to Lewis.

***

"I can give you instant beds," Lewis said.

"Instant?"

"Only kind I got."

We followed him to the engineering building.

"4-D printing."

"Four?" I said.

"Old technology. Been around a long time but we haven't had occasion to use it. How many beds and what kind?"

"Basic, single hospital beds. Twenty, I guess. How you going to do that?"

Ask Lewis and you will get an answer.

"We talked about this in class, you don't remember?" (Always our teacher.) "Integrated 3-D. Smart-polymer, shape-memory materials with different responses to heat or water, the energy source." We didn't answer. "And...?" still teaching.

He waited.

"They start to fold?" I ventured.

"Yep. We'll bring you the flats this evening. You'll have your beds tonight."

You have to love science.

As an afterthought, Lewis said, "I heard they rioted." He didn't see it, deep in his lab work.

I said, "Yeah. The bad news freaked them out like you said it would, and they're in the infirmary. It doesn't look like they'll be leaving anytime soon, so. They may even need to expand the infirmary."

He thought a minute, he understood how hard this was for settlers, then he said, "Let me know."

"You'll be the first."

***

The next morning was the day of Rob's dad's burial. The sign business and the rioting and the overcrowding of the infirmary, that all happened over a day and a night during which Rob's dad was frozen in our sub-zero atmosphere, the least of the settlers' worries. There would be a memorial. A grave dug outside the colony with pickaxes, was ready.

"I don't want to be there," Rob said to Lewis. "I can't stand to see him put under permafrost. Why do they want to do that? I don't understand. Out in space he'd be, you know, part of the universe again."

"He'll be part of the universe under permafrost," Lewis said, which is true. "The parents want to do something for him. They want to talk about him and say nice things. He was a nice guy, everybody liked him."

Rob gulped, stifle emotion, and told Lewis he didn't want to hear the nice things people said and go through those painful feelings again. Instead, he said he wanted to put on jetpacks and get away from the whole burial thing.

***

We flew low and high, did some loops, and checked out our welcome sign which was looking good, exactly like a giant had stamped it on the rusty regolith, the blowers keeping it clear.

Rob felt better. We did formation flying that we'd never done before, then tried something else new; held hands and did twirling acrobatics. In memory pieces we'd seen skydivers hook up and do acrobatic moves before opening their chutes but our moves were better than theirs could possibly be because we weren't falling and had more time and total control.

Our jet heat release always makes a faint vapor trail so we were leaving nice designs in the sky. Through them, from a certain angle, you could see the two moons, Phobos and tiny Deimos.

"That's so cool," Rob said.

When I saw through his face glass that his eyes were tearing, mine teared too. This is a good time to explain how close we are.

Sex is pleasant here, way more pleasant than in strong Earth-g, we're told. No heavy lifting they say which I don't entirely understand. Girls are inoculated young so we won't get pregnant until we want to, free love the parents call it. It seems normal to us. The inoculations didn't do well on the adult settler women but on us, the indigenous girls, they work fine. We Gen2s pair off very young, as friends, and experiment with sex whenever we want. I'm still inoculated and Rob and I have been together since we can remember. We were the first two born here, playing

56

together, sexual together really young, I don't remember when.

I can't imagine having sex with another man for the first time after I'm grown. Being naked with someone for the first time as an adult seems just awkward and difficult and it explains a lot about scenes in films and television because you don't know that person. Not really. So it's no wonder to us that pairing off on Earth is hard to get right.

That day, tears in my eyes when they were in Rob's was totally natural. We feel pretty much what the other feels.

"Want to go down and get the parents to look?" I said. "It's interesting to see the moons like that."

"My mom won't know," he said.

Of course. And my mother was not well in the infirmary with my dad who insisted on staying there because "she needs me more than ever." So we let it go.

The vapor trails didn't last very long anyway, and really, parents were only interested in the Earth's moon which we Gen2s have to agree is pretty impressive. Huge. Settlers always boast about it being big and bright with a permanent international station and its own reconnaissance orbiter and "we go there all the time."

We continued to jet around while Rob's dad was placed permanently under regolith.

13

"It's not at all like the plague," said Chang.

Rob had done some reading trying to find out what caused the parent breakdowns. They were comfortably accommodated in the infirmary but we needed more than ever to find out what was wrong with them.

"Plagues wiped out whole populations on Earth," Rob said.

Both doctors shook their head.

"This isn't a disease," said Sylva. "It won't kill them. Nothing's wrong with their organs. It's in their heads, their brains. It seems like just being here caused it...morbid homesickness."

"It's definitely not physical," Chang said. "They're in good health."

"Listen to this," said Sylva. (You can omit the following, NASA, but we found it an important key to understanding what happened to our parents.)

"Back in the sixties," she said, "some Cambodian women were brought to the U.S. after the genocide in their country. They couldn't speak or read English so they couldn't take a bus and go anywhere. They couldn't recognize American food or money or anything about our culture and they started going blind. Thirty or more of them were eventually declared psychosomatically blind, no sight whatsoever. I think there's something similar going on here."

I said, "You mean being here is so hard, their brains...?"

"...have taken over because their bodies are overwhelmed. I think the settlers are like those women, only the women didn't get irritable and reckless and belligerent. They went blind, their brain's choice for protection. For your parents, it's too hard to handle their feelings so their brain's protection is making them act out. And it works. The louder and stronger they express their frustration, the better they feel. For a time anyway."

Rob said, "Did those women ever get their sight back?"

"They did. They were taught the language and learned about our food and money and buses and they integrated socially. Gradually they could see again. Their brains let them take a look."

That sank in and I said, "I hate our parents feeling that bad."

The four of us talked about the attention that needed to be paid and came up with almost nothing, just keep them comfortable and distracted which the memory pieces were doing.

Before we left, Sylva said, "Whatever this is, Rob, you have to remember, your dad was an accident. He was only trying to save your mother."

"I know that."

I said to Sylva privately, "If our parents don't get better, won't accidents like out the airlock door, keep happening?"

"To be honest, I don't know."

Outside with Rob, I said, "So what's causing it?"

"Being on Mars, right?"

"Yeah but..." We walked for a few minutes.

Rob said to me, "Are you smiling?"

"Not all parents have the problem even though they're all living this same Martian lifestyle."

"That just makes it harder to figure out."

"No, it doesn't. It doesn't." I was still smiling. "C'mon."

***

"Look at it," I said. We were circling above the grower dome in jetpacks. "Some parents are still working there at least part time, right?"

"Somebody has to."

I looped down across the big dome and Rob followed.

"It's a nice place to work, isn't it?" he said. "All those plants and moisture in the air. What could hurt them?"

"I don't know but your dad didn't work there, did he?"

"Only in engineering."

"Your mom did, right? And she's sick."

"Yeah."

"It seems like parents who worked in the grower dome are the ones having the worst trouble."

"Is that true? Do you know that for a fact?"

"Well, Lewis doesn't work there. Neither do Derek's parents

or Rich's or my dad. But Mom was in the bio all day and she's in bad shape. Dad's only in the infirmary to support her. He always worked in engineering, Rob, and everybody else in engineering is fine too. So is everybody in the council. That's pretty conclusive, don't you think?"

We hovered upright looking at each other. Rob said, "You going to check this out?"

14

"What gave you such an idea?" Mr. Thomas said.

I was sitting in his office at the council module, doing my due diligence.

"Because not all parents are having the problem," I said. "Are you?"

"Having the problem? No." He scoffed, proud. No breakdown for him.

"And you never worked in the grower dome, see? So I thought..."

He interrupted. "I work right here all the time. We download research every day to NASA/JPL..."

"I know that, but..."

"...yield per dome-acre, pollination progress..."

"I know that, and that's my point. You have to be here, not with the growers."

"Of course."

He was taking my visit as an affront.

"All the people who worked in the grower dome are sick," I said.

"Yes, but we're on top of that. We've got others there working an extra hour. We've got it under control, everything is taken care of for now."

He waited with a testy look and wasn't going to be any help, so I thanked him and left.

I contacted Rob.

"It not under control. The dome needs a full compliment of workers or plants won't survive and neither will we...what's he thinking?"

"What are you going to do?"

I went straight to the dome and started collecting vials of everything there for Lewis to test.

I took samples of the water in the hydroponic basins that contained the nutrient solution. I took samples of the plants and of the air only the growers breathed in that beautiful closed system: carbon dioxide for the plants from us, and oxygen for us from the plants.

I took samples of it all, then took the vials to Lewis.

***

"You seem pretty sure," Lewis said to me.

"Only growers are affected. What would you think?"

"That we should run some tests."

Which is what he did that same day and the results showed nothing toxic, nothing abnormal. I went wandering back to my module. Rob was there, waiting for me outside.

"What happened," he said. "You look awful." He gave me a hug.

"The tests showed nothing."

"Ah, don't worry," he said. "Try not to."

"I still think that's where it's coming from."

"Okay, but there's nothing more you can do about it today, is there?"

I shook my head, still discouraged.

"You know what tonight is, don't you?" he said. I didn't know what he meant. "It's Oscar night."

Oh, geez.

15

Oscar night was big for the parents, especially this year considering what they were going through and how badly they needed distraction. We needed it too. The event took place during our morning, was sent on to us in a data dump to be replayed in our evening hours, more like on Earth.

Rob and I went early to visit his mother, the second time that day for him. She was still in a state, unaware of anything. Rob always talked to her, told her all about what was going on as though she was hearing every word.

"She looks like she's waiting for Dad to come back," Rob said to me later.

"Does she ever speak?"

Dolf standing near us said, "She never says anything. We don't think she can hear."

Dolf's a settler, a nurse, and the biggest, strongest man I ever saw.

"You've always worked in the infirmary, Dolf, right?" I said.

He nodded.

"You feeling fine?"

"Oh, yeah."

Rob and I exchanged a glance.

That night there were added chairs for the visiting kids and the staff. We sat with Rob's mom, me holding her hand, Rob's arm around her shoulders. She looked almost normal.

Leaning close, I whispered, "Look around. Nobody working here got sick, not a single doctor or medic or nurse is cranky and breaking down. It's only growers that worked in the bio since the beginning, long time exposure to...something."

"What about the settlers who came later?"

"They're afflicted if they worked in the bio, but not nearly as much. It's gotta be coming from the bio."

Rob nodded about to say something, and the screens flashed

on, all of them, above us and all around.

There was applause. The parents had been looking forward to this for weeks.

Rob stayed with his mother. I went to my parents and watched sitting between their beds.

The 2055 Academy Awards show was announced--gorgeous color, spectacular dancing, singing, jokes and applause. Very entertaining. But not distracting enough to keep my thoughts off the grower dome.

I looked around wondering, could I sneak out, go there in the dark and nose around? No, I couldn't. Like the other Gen2s, I sat with my parents who explained what everything meant, some of it so current it was unfamiliar even to them.

It was glamorous, that's for sure, but the humor was incomprehensible. Much of it had to do with sex but in such an odd way we couldn't get it. So to please our ailing parents some Gen2s (including me) were fake laughing. There were jokes about famous people in the audience that got big laughs, even from the famous people. But we didn't know the people or what the jokes referred to so more fake laughing.

The actors and audience were dressed differently, something we don't see here. They looked costumed, the men all wearing black, the women baring a lot of skin. I kept telling myself it's not cold, they're breathing oxygen, but not wearing protection gear looked very dangerous.

Rob joined me and said, "You look disturbed."

"All the skin. Out in the open."

Another woman was bare her to her navel and when she turned, she was bare to the small of her back.

"Look at that," I said.

The dress was mostly discreet strips of sleek fabric.

Rob said, "I saw a close-up on some guy's colorful socks."

"How'd I miss that?"

That's how it went, watching the show amazed at things that were foreign to us but were totally familiar to our parents and much longed for, not a moody face among them.

I stopped trying to understand the show and studied the happy moms and dads. They had all, without exception, worked in the grower dome, every single one of them.

"And the Oscar goes to...?"

***

Next day in class I was watching through a port for speedy little Phobos but my mind was on the bio as the cause of the problem. I was certain of it. Phobos was supposed to show up in another few minutes. Deimos too, our other moon, but it's slower and so tiny it looks like a star. Phobos is my favorite and I was going to write about it. My tablet's readout was ticking down to Phobos arrival time.

But as I watched and wondered what could be toxic in our grower dome, my eyes focused on dust buildup on the flat port.

That wasn't normal. It's never good.

There's always some atmospheric movement and some dust, but not buildup on something upright and flat. The constant crawl on our screens didn't warn about wind. But wait, our wall screen feeds were turned off so we could use them for homework.

I kept watching my tablet and right on schedule, there was Phobos moving into view. My plan was to go out, make some Phobos notes, then go on over to the council for data I could use to prove my idea.

Nevermind the dust buildup.

I pulled on my skin, locked down my face glass, and ten steps out the airlock door the sirens went off with a blaring audio warning: CHECK YOUR SCREENS.

I didn't need to check a screen.

A wall of dust hundreds of meters high was rolling in over the top of the domes, higher than any I'd ever seen and moving toward us like a giant boiling wave. One you can't surf. It spread across half the horizon. We're used to local dust storms, those are frequent and harmless. But this storm was huge and no way was it harmless.

Inside the class room, Rob was looking for me. The kids were told they couldn't leave, they had to stay until the storm passed, doors locked.

A message flashed on every wall in every structure as well as inside my face-glass:

DO NOT LEAVE YOUR MODULE.

MAJOR DUST STORM. HIGH WIND.

Our satellites care for us like nannies, but this time Nanny failed. Gave us a late warning and only last minute data.

Wind: 300 kph. (186 mph, Earth readers.)

Advisory: WIND INCREASING. STAY INDOORS.

KEEP IN TOUCH WITH COUNCIL.

Good Nanny doesn't want us out of her sight because that last announcement sounded like trouble was expected and council wanted to know where we were at all times.

It's just wind, I kept thinking. Wind and dust.

These storms happen on Earth too, only with sand. But our atmosphere is so thin and the dust is so fine it isn't harsh like on Earth. The molecules aren't close enough together to make a wind thick enough to do any harm.

Dad explained it to my New Jersey aunt: "It's like water from a sprinkling can and water from a bucket. One hardly bothers you. The other knocks you down. And they're both falling on you at the same speed."

Enough explaining. Our talcum powder dust blinded me and I decided to sprint to the council, it was close. But too late. I couldn't see my hand in front of my face, literally. I held out my arm, couldn't see my hand. The dust was enclosing me like a blanket.

It was like a white-out in snow on Earth. Dad's told me all about that kind of thing so I know it's the same. There is nothing for the parietal lobe to relate to, proprioception is impossible. Are you with me on this? Without that, my brain couldn't help me move.

I dropped down, spread out, and closed my eyes, gravity my only guide. Our colony's structures are sealed, smooth and strong against high winds and blinding dust storms not to mention the constant interior pressure of 1 atmosphere. But I wasn't smooth and strong, even in my sealed skin and glass. This was the first time I'd been caught in a storm and lying on the ground in Mars-g was the only sensation my brain could relate to.

My heart was thumping. I couldn't stay there. The dust could be this dense for days. But I didn't dare get up. I balled myself together like the armadillo I'd seen in David Attenborough memory pieces.

The wind got stronger and I knew I should get inside. I raised to a crouch and feeling reasonably stable on my hands and knees went head first into the wind. I didn't know it, but I was headed back toward the school and Rob was looking for me. He didn't know when I left and only missed me when the warning came on the screens. He was peering out a port but couldn't see through the storm any better than I could. Lewis was busy keeping the kids happy so Rob pulled on gear and came out to find me.

He had to drop to the ground like I did, dust blinded. He called for me and his voice in my headgear was garbled. It didn't help except to tell me somebody was trying to talk to me and knew I was out there. That was comforting but there was no way to know which direction the voice came from. I struggled to stand in the blinding dust. Impossible. Even my balance was impaired.

"Rob! Keep talking to me!" I said. "Rob!"

Garble.

I still couldn't make out the words and had to idea the direction they came from.

GPS was gone too. That's right, a storm blocks out our satellites. No more triangulation.

I kept calling and getting garble in return. Rob was getting only my garble. The noise of the wind was increasing and in our thin atmosphere where so little can be heard, that big sound made my heart thump harder.

Then we saw each other, blurred images through the dust only a couple of meters apart. But in that instant, a gust caught us both and pushed me back and Rob sideways. We didn't know where the other had gone.

Another gust hit and pushed me. I lay flat as I could for a long time, probably only a minute or two. I tried to remember how long these storms last. They go on for hours, sometimes days.

Where's Rob?

Where am I?

Where's a hab or dome or any colony structure?

Very disturbing sensations...then a bump.

I swept the dust off my face glass; Rob was on his hands and knees in front of me. We gripped on to each other and I saw his lips moving but couldn't make out the garble. I shook my head and shrugged. He nodded.

Okay, no communication.

He seemed to know where we were and he held my arm and we snaked (more Attenborough) along the ground. At one point the wind let up a little but the dust was just as dense and it didn't mean the storm was weakening. My brain settled down and started working. I remembered.

Some tricky dust storms have an eye like hurricanes. One side can pass over you, then you're hit by the other side with dust coming from the opposite direction. But the let up we felt in the wind was brief, only a few seconds. It wasn't an eye.

We snaked along until we came to a dark mass rising in front of us. The infirmary.

As we got closer we saw an airlock door open. Somebody was coming out wearing no gear. Another figure, a nurse, was trying to pull the person back in. We made out that it was two people the nurse was trying to pull in--fuzzy images. But the nurse wasn't anywhere near strong enough and the two unprotected figures pulled away. The nurse fell back and the airlock door slid shut.

The two figures were out without gear, without oxygen, holding on to each other in sub-zero cold. We started toward them but it took a lot of squirming to reach them and when we did they were crumpled on the ground embracing each other, the wind gently rocking them. The dust and sand were already building up against their bodies, no gear, no oxygen, no protection whatsoever--like Rob's dad, wearing only indoor body thermals, both of them, dead.

16

"It's what they wanted," said Dr. Sylva.

The deceased Bianchis lay together in what looked like peaceful repose. Rob and I were exhausted after managing to get the two bodies in the air lock.

"A lot of them want it now," said Chang.

I looked at Rob, his head lowered, thinking of his dad, of course.

Chang said, "We've got to get these two in another room without being seen."

He called a medic to bring a gurney. The medic arrived and said the parents didn't know anything about the deaths.

"A few are watching the storm through ports but they can't see anything and the others are watching something on their screens."

"Good," said Chang. "How do we roll them through?"

Silva said, "I'll take care of this."

She left the airlock and went to one side of the big infirmary room.

"Come over here everybody," she said. "We need a vote."

They could all walk, you understand, they just preferred lounging on their beds. So they gathered around her, interested. What could she have in mind?

"What do you want for a snack?" she said. "Popcorn?"

There was a lot of interest in that. Our parents were more childlike every day.

"Raise your hands...hold them up so I can count."

She had their attention.

"There's an alternative," she said.

"Tell us."

"Peanut butter cookies, but you'll have to wait for those."

That caused a happy discussion and different hands went up. Others came down.

"Let me count for cookies," she said. "Keep them up so I can count."

With all eyes on her, we rolled the Bianchis through behind their backs.

In the examination room, the Bianchis still covered, Chang looked around and said, "Well, they're all going to want this," and shook his head. "We're gonna need a bigger boat."

I'd heard that before, knew what it meant though I never knew the source, and yeah, they'd need a bigger room for the dead.

"They're not getting any better," he said. "And this, what the Bianchis did, is such an easy out. They don't want to recover." I was frowning, didn't want to believe him. "They told me they want to stay the way they are, Kit. Don't want be rehabbed. They like being pleasantly blurred, that's what they're calling it."

"Are you sure? I can't believe it. What do they do all day?"

"Old films. Sports. Games they can remember play by play. Memory pieces of course, lots of those. Some watch all day and then fall asleep. It's as close as they can get to home. You can't blame them."

"I don't blame them. "I want to...help them."

"They're not uncomfortable."

Parents were talking to each other, some still in bed, some playing cards and board games, most of them watching something on a screen, hardly noticing the storm and still unaware of the Bianchis.

"Waiting for cookies," Chang said. "Happy. They don't want to live in the colony again. They don't want to see the colony again except maybe to do what the Bianchis did. When they find out, it might even encourage them."

"Well the rest of the colony isn't happy," I said. "We've lost over a third of our workforce. Thomas says it's all under control but it isn't. There aren't enough in the biosphere now to keep it going."

Rob said, "What keeps the others from doing what the

Bianchis did?"

"I have no idea," Chang said.

I whispered to Rob, "Want to leave?" Thinking of his dad, this had to be painful for him.

"Not in the storm."

We tried to come up with solutions as the storm raged, shaping up like the longest dust storm in our twenty-year history. The scroll on all walls was telling us visibility was still near zero.

DO NOT GO OUT EXCEPT IN EMERGENCY.

Rob got a text from Lewis and showed it to me: NASA says the storm is covering almost the entire planet. Looks like it could be a month before it clears.

"A month?" I said. "Is he serious?"

"Lewis doesn't joke, neither does NASA."

A fresh advisory scrolled across all screens.

NO OUTSIDE MOVEMENT.

CONTACT COUNCIL FOR EMERGENCY NEEDS.

17

It turned into a very long day.

Class was canceled until the storm cleared and parents were contacted. We had to get the kids back home.

How could we do that, dust blinded?

We gave it some thought and came up with this: a long rod.

While still well oriented just outside the airlock door, we laid it down pointing in the direction we wanted to go. Hovering over it in a low lizard-like crawl (Attenborough) we got to the end of the rod, moved the rod ahead in the same direction as the trace it left under us which we'd kept from being covered with dust by our lizard crawl. Very tiring to arms and legs but the rod kept us moving straight toward our target.

Rob and I managed to crawl with kids tethered behind us, one behind the other like ducklings. They used any crawl they wanted, giggling, a big adventure. We got them to the homes of parents still living in their habs. Children with both parents in the infirmary were led to the homes of friends where they were staying. The big adventure for the kids was a strenuous business for Rob and me.

We got back to the infirmary and sank onto couches, breathing hard.

"A month long storm?" Rob said when we could talk again. "Impossible."

"It's Mars," said Chang. "It happens."

I contacted Lewis.

"Lewis says to wait till morning then stretch NASA tether waist high between all the structures so we have something to hold onto."

"That'll work," Rob said.

Arched glass hallways connected some structure but not all of them. Mega strength NASA tether was needed everywhere else.

Of course, Lewis and Rob and I would do that. But it was late and we were beat.

Silva said, "Don't go out again. There's a bed in the medication room."

It was narrow but we hardly noticed. Rob and I lay down, reached inside each other's thermal, and dropped off. We woke next morning in the same position. So, early morning, well rested sex, about the best. Especially Rob holding me and telling me we'd be all right as though we were in great danger which we knew we weren't. (Sort of like scenes in Earth movies.)

A high protein infirmary breakfast and we were ready to go despite the continuing high winds and low visibility. (Lewis will do a separate, data detailed report on all we learned about the storm later, NASA. You can expect it.)

We spooled out tether from module to module, brushing dust off our face glass every few minutes.

There was a joke going around the infirmary: "Badminton, anyone?"

Hilarious they thought--high winds, feathered shuttlecock. Some parents teared they laughed so hard, a strong indication they were far from recovered. When we ran tether to the infirmary we could see inside through the ports. The isolation had created a party atmosphere. The patients were decorating the place and beckoned us in.

Soon as I stepped inside Zelda and Ben's mother said, "Hold

this," a crepe paper strip.

She lifted the other end to the top of a port and attached it. She took my end and attached it at an inside door. The interior of the infirmary had become their home away from home, away from modules, colorful and much more Earth-like. It was crisscrossed and bannered with decorations for birthdays and holidays, Fourth of July very well represented, of course. I saw Rob's mom still sitting quietly, not watching, staring off. My own parents were playing cards; competition level bridge, their favorite.

Dr. Chang saw me and nodded. His patients were happy and occupied. As I looked over the room full of parents and examined their faces, my idea about the cause of their affliction came back stronger than ever. The storm did not wipe it out. These parents were without exception original, grower dome workers.

***

Using the tethers and calling myself unstoppable, I made my way to the council. There they have a record of the entire Mars population, every piece of information about every person in the colony including their original Mars One application and a full background history of their work activities since arriving.

What I wanted was simple. On Mars, who worked where, when, and for how long? Here's what I learned:

1. Most bio workers were settlers and had been there for years, learned what to do, enjoyed it, and stayed until they were afflicted.

2. Only a few who came here in the second batch, worked there. They were afflicted but not as badly. They hadn't worked there as long.

3. A few others had worked in bio for some kind of research but only for a few months and weren't affected at all.

This absolutely confirmed my notion that no one got sick if they did not work in the biosphere for an extended period of time.

How easy was that to figure out?

Long time growers among the original settlers were severely affected and I now had documented proof.

I did cross referencing to be sure of my data, saved it on my tablet, and tether-walked to Lewis.

***

"You look smug," he said.

"Something in the grower dome messed with those workers and *only* those workers," I said.

"I tested samples..."

"I know, but that doesn't mean it's not there, just because we can't find it. There's something in the bio that's causing this, Lewis. Look, look..." I showed him my data. "Unless we want to build an addition on the infirmary we have to get the workers that are in the bio now, out. And I don't know who'll replace them."

"Who's there now?"

"Anybody who hasn't worked there yet and they're being contaminated. And there aren't enough of them, Lewis. Who else can we put there?"

"It takes years to affect them."

"No, it doesn't. Chang says it's happening faster. He thinks whatever it is it's stronger now."

"Yeah? Well..." He thought a minute. "We need a rotation chart."

Yep. Next best thing to a cure.

We repositioned everybody's work assignments including mine and Rob's. Lewis had to stay in science. So did Derek, his assistant, and Zelda because they're so important to our survival. Rob and I would work with the growers, right in the center of whatever it was that we couldn't detect. We'd be in that perfect, deadly, sweet spot among the veggies and miniature nut trees.

"We'll find it," I said.

***

Ben overheard me talking to Zelda about the shifts in work assignments.

"I want to work there too," he said. "I want to work with the

80

growers."

"You're too young," Zelda said.

She was now his mother with both their parents in the infirmary.

"I can do it. I know how."

He meant the pollinating, and of course he could. His small hands could handle the vibrating wand better than adult hands.

"But it could hurt you."

"How? I'll be careful."

We explained how long term it was dangerous.

"But I won't be in there long enough to get hurt."

He was right about that. Our rotation chart was amended. We agreed to let him and a few other kids work scattered hours throughout the week but never for extended periods and never accumulating much time.

18

Gen2s, even young ones, along with Lewis were now running things. That news went back to NASA/JPL along with our redistribution of work assignments.

> JPL: Kids are running things?
> MARS: Yes.
> JPL: Cool.
> MARS: Not cool, Jeff. Life threatening.

JPL: What is it exactly?

MARS: We don't know. That's the problem.

JPL: Give us something to go on. Tell us more about their behavior. We have behaviorists that will tear it apart. They'll figure it out.

MARS: Impaired judgment. Belligerence. Reckless behavior. Dividing into rival factions. Depressed. Suicidal.

Our failure is dreaded because the Mars mission is so important to NASA and space exploration in the future. If we died so would NASA's budget. But as bad as losing funds is, it wasn't nearly the impetus for finding out what had gone wrong that our looming death was.

So I said, "Are they taking us seriously?"

"Sounds like it," Lewis said. "But don't forget they're focusing on two new planets that look better than Mars for terraforming."

"Yeah, well...they're not in the solar system. They'll need a wormhole to use them."

I put that information out of my mind.

JPL knowing made it official and kind of overwhelming. Lewis, Rob and I were in charge now, directing, managing, setting new rules. We called a general assembly of colony members, the ones not in the infirmary, the "still working". We met in the main council chamber.

Colony members were expecting the usual report of news from Earth and updates on projects and possibly an explanation of what was ailing the other settlers. But there was none of that.

"Today, our assignments in the colony are the subject of the meeting," Thomas announced, and immediately turned the meeting over to Lewis.

"First, there are some safety reminders," Lewis said. "As you know, the storm looks like it will go on for another week or so even though it's much lighter, and we can see better. So, number one," he held up a finger. "Please continue using the tether leads even as visibility increases. A gust could blow you down and you'd be disoriented, might get lost."

There was scattered talk. They understood.

"Number two," two fingers. "For your further protection, we're passing out new work assignments."

There were murmurs in the group as Rob and I passed out the pages. As they read they started talking. Their babble increased. They weren't happy.

"It's nothing drastic," Lewis said. "We just want you to be acquainted with the work everybody does."

A woman's voice said, "I like what I do at council. I don't want to move."

Another voice said, "I don't want to move out of the bio. I just got there. Why are we doing this?"

There was more talking, louder, and Lewis interrupted.

"It's necessary," he said. "This is absolutely necessary. You know how many are in the infirmary and until we find out what caused their problem and get them back on their feet, we need to replace them. We all have to be able to do the work here, all of it. We can't live without growers but we'll survive with fewer council workers. The specialists, the engineers and builders in the printing module and the infirmary staff can't be moved which means the rest of us have to share the load."

A woman's voice said, "You have my daughter's name on here. She's twelve."

"It's only a couple of hours a week," Lewis answered. "Our kids need to know how to do the work too."

Not a word about the bio possibly being the location of the problem.

The gravity of our situation started sinking in and they went quiet. Then after a minute they started talking again. At the same time, the wind and dust kicked up and made a boom against the council dome which carried very well inside. That was no help.

I stepped up and said, "That's it for this evening, everybody. The new assignments begin Monday and we need you to cooperate until we can get back to normal." They groaned. "Be on time and keep up the good work." They began to stand. "And use the tethers."

There were no more questions but they weren't happy. They didn't like the change. The kids of course would be very happy with it, working with adults at adult jobs. Wow.

***

The month-long storm wasn't a month. It was less than three weeks. JPL peppered us with dust storm questions the whole time. Their satellites couldn't see through it. We had to update them.

> MARS: Two weeks of storm and visibility is much better, winds weaker. We're still using tether lines to walk between habs.
> NASA: Be careful.

So we continued to use passages to get from building to building when we could. We missed our usual beautiful views those weeks, our biggest weather complaint. And the satellites couldn't see a thing except dust cover. Meanwhile, we were back to normal in a new-normal way.

Ex-council workers and a few kids were in the grower dome. The unbalanced parents were dreamy and reasonably happy in the infirmary. Everybody else was at their various jobs.

The ball field remained busy but with young kids. No more fist fights. But parents who hadn't been working as growers and were not sick like those in the infirmary were nevertheless beginning to cave too. In a different way. They weren't as belligerent, and their judgment wasn't as impaired as the original growers, but something was wrong with them. They really did seem homesick. Yet they hadn't worked in the bio.

19

"I'm gonna explode if I don't..."

Next day working with Rob in the grower dome I was itchy and anxious. We were pollinating tomato plants. Ben was working not far from us, his short, weekend shift.

"And it kills me to see him here. We've got to find out..."

A help monitor walked by.

"Need anything?"

"No. Think we've got it," Rob said. We went back to work.

"Listen," I said. "There's something I have to do during our break."

What we did was to stand near the dome's edge where the glass and spars meet the concrete base so we could look out over the entire area. It's almost

0 acres. We're told is like the sport field superdomes we'd seen in memory pieces. Ours was built large to accommodate more food for more settlers still to come.

"Okay," I said to Rob. "What do we see that's different?"

"Everything. It's bigger than the other domes. It's the only one full of plants and water. It has more workers."

"So...what's happening to them here?"

We examined the spars, the glass, the hydroponic basins, everything Lewis had tested. We walked through the dome looking for anything unusual. Last of all, I took a good, long look at the concrete.

"It was made on Mars with Martian regolith, right?" I said.

"Right. Like in all the structures."

"Yeah, but the others don't have the CO_2/oxygen cycle."

"Right."

I looked around. Nobody watching. I scraped at one of the concrete bases.

"What are you doing, Kit?"

"Getting a sample. Thomas better not see me doing this. It'll kill him."

"He's not working at council any longer, don't worry."

It was hard to get the concrete to give up crumbs. Rob stood with his back to the workers who weren't paying attention to us anyway, and worked at it with a little tool we carry, a kind of Martian Swiss knife, clever people those Swiss. I doubt it was made to stab at base concrete in a dome on Mars but it did the job. He made a small chip, then managed to get a handful of powdered crumbs which he gave me.

"What are you thinking?" he said.

"That they couldn't have made the same mistake twice but the symptoms are the same."

"What are you talking about?"

***

With Lewis an hour later: "Not very scientific," I said. "My sweat's mixed in but..."

"What do you want me to do with it?"

He was looking at the sticky crumbs that I was shaking into his hand.

"Reverse engineer it."

"Why?"

"Maybe this is what's making them sick."

"Concrete's at the base of all the structures..."

"...I know, but the biosphere on Earth, remember? Decades ago?"

He thought a minute, remembered, and nodded.

"Please test it," I said. "Will you? Now?"

He tipped his head, skeptical, but he turned away which meant yes, and I went back to my hab still jumpy with energy.

While Lewis reverse engineered concrete crumbs there were two more deaths.

20

"Embracing, lying right outside the door," Chang said to me. "Like the Bianchis. They're romantic heroes now."

"For couples, you mean?"

"Yeah."

"What do we do?"

Chang, calm and efficient, said, "Put a guard at the door but I don't have extra medics for that and it'll have to be 'round the clock. That means losing a medic or nurse each shift. And it has to be somebody strong enough to stop two people."

There was no immediate answer. I offered to do it.

"You're not hefty enough, Kit. You're more useful working with Lewis and Rob on why this is happening. You are working on it, aren't you?"

"Oh, yeah. I was just with Lewis and...nevermind."

I looked back at the airlock door. Bear-size Dolf was there.

"He's our best," Chang said. "I don't like losing him on any shift."

"This is ridiculous," I said. "I'll find somebody."

Looking around I saw very few to choose from. Then I saw blind Mr. Elkins and remembered him speeding toward us on his mono with a red dust rooster tail.

"Mr. Elkins...?" walking toward him.

He was soon at the door with his short, solid stick, the one he uses for echolocation. It works perfectly in our close quarters with solid-form floors and walls and 1 atmosphere. He walks around a room making little taps you hardly notice and you wouldn't know he's blind. He was a perfect guard and he wasn't sick like the others. (Never worked at the bio!) He'd been helping with small tasks and now he was given a big one. A life-saving one.

"He'll know when someone or something is coming near," I said.

I got a skeptical look from Chang. But Elkins was at the door, armed with his stout stick and confident.

Note: Despite the disappointment in losing a Mars One original, Mr. Elkins has turned out to be one of our most valuable assets. Infirmary airlock guard is only one.

"It's good to feel useful, Kit," Elkins said to me. "Know what I mean?"

"I do, I do."

"I'm insomniac," he said. "So I'll do night shifts and a day shift too."

We watched and after a while, Chang felt we had to test him. We walked toward Elkins making no sound (we thought), him tapping occasionally, and when we were still several meters away he stood, stick raised and ready.

"Who is it!"

"It's me and Kit," Chang said. "What if we were a couple up to something suspicious and we rushed you?"

"Side of the neck with my stick." He made the gesture. "I'm trained you know."

I didn't know what he meant, some Earth training maybe?

90

"This is going to be fun," he said and sat down smiling and ready.

All right, then. The infirmary airlock had a guard.

Next morning the parents discovered they couldn't get out and were furious. They felt incarcerated and wouldn't have it. They politely refused to eat.

"No, thanks."

"Thank you, no."

"None for me."

That's how it went, parents on a hunger strike. Not a strike to gain something unless you call wanting to die a victory which I guess they did. I couldn't blame them.

I went to my mom and dad. "You're not eating?"

"No. Everybody's agreed. Solidarity."

They smiled at me, serene. They didn't want to step outside and die, I'm sure of it, but they were solidly behind the others who did. It's so strange looking back on that, everybody determined and calm, nobody uncomfortably hungry yet.

I sat on Mom's bed. "This is going to feel unpleasant, you know."

"That's all right."

I was angry but I nodded and went to Chang.

"A hunger strike? What do we do?"

"You tell me," he said. "What will get them moving? Their bodies are not debilitated, not yet."

"They're still strong?"

"Yeah. But not for long lying here melancholy and starving themselves."

I gave him my I-know-what-to-do smile. I admit it, I'm smug sometimes.

21

A huge trampoline was set up in the baseball field. Those not on the trampoline could sit in the bleachers and watch.

The trampoline was meant to be used by all the Mars One settlers, but it never was. No one was interested. But I knew about it, and after being stored in the back of the lab for years it had no deterioration that Lewis could find (no Martian dampness), so it was set up and the parents were bouncing on it, having a great time. Nobody could be hurt, not possible, not enough Mars-g for that. And only a soft landing if they hit the edge of the frame. It was modified with thick padding sloping to the ground.

Chang and I watched.

"I could never get them interested," he said.

"Well, they are now. And on empty stomachs."

The hunger strike wasn't on the parents' minds. They were having too much fun rising up and up and coming down slowly, beautifully, unhurt. It looked like normal speed to me but to them, they were doing something that on Earth was much harder and faster and certainly more dangerous. For once, to the parents, Mars was better than Earth.

"Perfect," Chang said grinning. "Just what they need."

My own parents were watching the fun, not yet on the trampoline. I'd arranged that.

"They'll be the special event," I told Sylva. "When the others are pretty much exhausted."

"Whatever you say," she said.

Like Chang, she liked what she was seeing and trusted my idea.

Finally, we called everybody off the trampoline and sent them into the bleachers for the big show.

A little background: Dad taught me to dust surf because he'd learned on snow which is much harder and not very forgiving. Falling on hard pack snow in strong Earth-g has got to be wicked. He was good at all sports but it was on the trampoline that he was a champion which not many parents knew. He doesn't brag. And there's no call for trampoline champs in space. But his athletic ability, which he rarely mentioned, and his accomplished engineering studies which were well known, had gotten him into Mars One. It was the same pattern with my mother only she was a biologist/athlete. Now, both their athletic abilities were going to pay off. Even here. Especially here.

Dad leaped up on the trampoline mat, knew what he was doing, not tumbling and jumping around like the others.

He stood straight, focused, and started warming up slowly, bending his knees, pushing up, gaining height, raising his arms for more height which he did to spectacular elevation. He was going so high it looked like he might miss the trampoline when he came down but he didn't.

He was back in the athlete's zone again after all these years.

"I knew exactly what I was doing," he told me later, shaking his head amazed at himself. "Muscle memory, I guess, all those hours and years of practice..."

He began his routine, double flips and pikes, then twists so beautiful against the sky, so graceful, so scarily high that we were afraid to breathe.

"Why haven't we seen this before?" Chang murmured to Sylva.

"He didn't have a note pinned to his space suit, Chris."

She never took her eyes off high-flying Dad.

I always knew how good my dad was although I'd never seen it except on a screen and I never had an occasion to mention it or call on it. After a spectacular show, he came down into a crouch that dissipated all the energy, and he dismounted like the champion he was and still is.

There was wild applause and cheering. They called out for more but he didn't climb on again. Instead, he escorted my mother onto the mat, the real show.

Everybody went silent.

(This may be too long for report material but I'm proud of my parents and how helpful they were in the recovery of the others. Delete it if you must but I hope you don't.)

We were silent. Only I knew how good she was though I never had much occasion to talk about her skills. I should have. My mom was the other star, the world champion. Yes, better than Dad and a biologist too. No wonder she was allowed to come to Mars. Both of them--athletic, disciplined, smart, skilled, and ready to take on something new. But this trampoline routine was not new to my mother or dad. Like him, she'd done it under the most demanding pressure.

She stepped on the mat, hands at her side and I knew I was going to see what world competition must have been like. She was concentrating. She crouched and pushed up, gained height, pushed again and again, raising her arms. She began getting height like Dad, even higher. I was scared for her. It looked like she really would lose control up that high. The parents watched, afraid to make a sound as she kept rising higher and then began her pikes and flips and shapes. That's what it's all about, how beautiful a perfect shape (arms and legs at competition designated angles) you can make with your body at a great height, this time higher than any competition ever on Earth.

She did what she could never do in Earth-g but on Mars that day it was child's play. Really, so easy. She was smiling and spinning and rolling, and always landing on the spot. Incredible.

Impossible. But she was doing it. And when she'd done everything, better and higher than ever before, she came down for the last time to a crouch and stopped. She jumped off the trampoline, smiled and raised her arms, a world champ, turning around, happy.

In the bleachers we went wild again, not just because of what she'd done but how she'd made us feel. Nostalgic parents were elated. They felt special seeing her do what had never been done unless it was on another planet or in another galaxy. I felt proud, puffed up with pride, I admit it. We were united, nearly every colonist feeling good for the others and not thinking about our problems. Entertained not by images on the screens that surround us, but by a real, living show.

Lewis wasn't there as usual. He and Derek were back at the lab hopefully reverse engineering my concrete crumbs.

The parents' happy talk went on inside the infirmary even after a big lunch that they chose not to lie down and strike, but to eat, sitting at their bedside tables laughing and talking with energy.

"Who can lie down after that?" I heard somebody say.

Dad was so proud of my mother he lifted her hand and led her around the infirmary like a thoroughbred. Everybody applauded again, for both of them. They were winners. Healers too for a while.

As they ate well, nothing more was said about a hunger strike though a lot was said about doing the trampoline again.

Rob said, "Good job, Kit."

I said, "Yeah," and couldn't stop smiling.

I was congratulating myself on the success of my trampoline plan and possibly the best day in the colony for a long time when Lewis walked in. He did not look happy.

22

"Not good, eh?" I said, outside, away from the jubilation.

"It's the concrete, all right."

"What's wrong with it?"

"It looked normal at first and then..." He stopped.

"Say it. We going to die?"

"Yeah, if we don't fix it. And I don't think it's fixable."

"Why?"

We walked toward the lab. "The concrete was cured wrong like in Bio 2, and it's absorbing more oxygen than the plants can replace."

"That caused the breakdown?"

"Yeah. It's still doing it to whoever works there and the rate's accelerating. Those Bio 2 people lived there 24/7. I'm surprised they lasted as long as they did. It's like working at 18,000 feet on Earth."

"I'm not sure what that means."

"Not enough oxygen. If you're not used to it, it weakens you, scrambles your brain."

"But we've got regulators and oxygenators everywhere."

"Not in the grower."

"So we add them."

"Won't work. Either humans cycling with plants is in place and works, or we use some entirely different system."

"Like oxygenators."

"No, no, the whole system has to change. We've got to ditch our Deepwater Culture System and put in a different one. Your mother knows about this. I really should consult her."

"No, no, no. Don't."

"She'll have to know some time."

"Okay...but...some new system? How do we..."

"I'm just saying. We could switch to Nutrient Film Technique. That's continuous flow with water-based nutrients. Plants take more oxygen from the air and the dome goes fully on oxygenators like all the others. Totally modified. No exchange cycle."

"But would it work?"

"It's Biology 101."

What a decades-long fruitless search didn't give us was now what we needed: available flowing water on Mars--not trickles, not melted dew droplets, but available, continuously flowing water. Rob joined us in the council building and Lewis explained the Bio 2 effect on our parents to Jeff at JPL.

***

MARS: They're down and out and we need flowing water for a new system. A major part of our workforce is lost and it's dangerous for the workers taking their place. The grower dome is our only source of food. No Safeway up here, guys. Your thoughts?

JPL: Good God, lots of thoughts. Nothing solid at the moment but give us some time. When did this happen?

MARS: It started when they got here. We discovered what caused it today. We've got things running but just barely and not for long. Ideas?

JPL asked several questions which Lewis answered. They were stunned to learn about the concrete.

JPL: Cured wrong, you're sure?

MARS: Like Biosphere 2 in Arizona. Same mistake. It's life threatening here. What do we do?

JPL; We need time.

MARS: Don't have much of that.

JPL: Right. Any special conditions we need to know about? Anything new there, atmosphere, topography, anything?

MARS: Nope. Just get it right.

Rob, Lewis, and I stood there thinking the same thing: After twenty years of successful living on Mars, the U.S. is the gold standard of space exploration and failure now would choke off NASA funds and suspend more flights. NASA was taking our situation very seriously and with settlers from ten countries living here, the world would take us seriously too if they knew.

I asked Lewis, "Do you think they know?"

"We're just a blip on their news. Their floods and weather are more than they can handle right now."

He was right.

Lewis said, "They're sweating bullets," a new Earth expression that I figured out.

"They should be," Rob said.

Lewis was looking away, deep in thought, not saying a word.

After a minute Rob said to him, "What?"

Lewis looked up and took a deep breath. "They're going to come up with nothing or something new that's useless." He was shaking his head. "We studied this a long time before we came here and..." JPL pinged again. "What band-aid fix is this?"

JPL: There's water in the permafrost at night that melts with perchlorate salt. It absorbs the frost and creates a layer of salty brine, The liquefied water sinks even lower. It starts to move and relocate.

MARS: Relocation does us no good. We need a strong flow that we can irrigate through our hydroponic basins. There's nothing like that here except under the ice cap. Is there?

JPL: There's potential for it.

MARS: You don't get it, Jeff. Even if we could reach that water we'd still have to get it here.

JPL: Right.

Lewis sounded scared and angry. Jeff seemed shaky. Rob was steady.

They messaged back and forth about water under the permafrost even at our latitude, but especially up toward the poles where there are likely springs and a deep, buried ocean. We couldn't reach any of that. With no good news Lewis signed off, still angry, and we left.

Walking back to the lab nobody said anything. Lewis was thinking. Rob too. I had faith in both of them and in myself. I knew as much as they did about unreachable Mars water--it's unreachable.

Rob and I went to his hab, so many habs empty with parents in the infirmary. We sat thinking. Being a scribe/sketcher I don't know as much science as he does, but I have good ideas. I'm inventive. I do know that.

I sat and sketched. Without a molten layer, Mars doesn't have a magnetic field...the other strata nearer the surface could have flowing water...possibly. Not much to go on.

Rob was studying old research.

The hab was quiet.

23

Next morning: "Want to road trip to the pole?"

Lewis was asking it seriously.

Rob said, "Why?"

"I talked to JPL again and the only idea they've come up with is an old joke. But I had the same thought so..."

"What's the joke?"

"How do you get to the water on Mars? Bomb it with an asteroid."

"Would that work?" Rob said.

"Maybe. Asteroids change anything they impact but it's hard to make it happen. And it's got to be exact. The odds are terrible."

"Is this a serious thing?"

"Only because there isn't a better idea."

"Haven't they been working on directing asteroids for years?" I said. "To keep one from hitting Earth?"

"Yeah, but it's all on paper."

"Have they run the numbers?"

"Yeah."

"So, should they try?"

"You sound eager," Rob said to me.

"I am. It's a last resort."

"Probably the only one," Lewis said. "So, the pole?"

I said, "I'm thinking, yeah. Do you want to, Rob?"

"What would we do?"

"Find a target, the coordinates of a place most likely to release water, a spring. They need something to aim at. MROs aren't as good as somebody standing there, so..."

He literally rubbed his hands together, eager for our cooperation. "Road trip, you two?"

A little later we were looking at Lewis's finger on a large map of the solar system projected on the lab screen.

"Right there," he said. "Mars, at the edge of the asteroid belt."

Rob and I are well aware of our location in space but our proximity to the asteroids was particularly impressive that day considering the plan.

"We're at the edge of hundreds of thousands of orbiting asteroids," Lewis said. "Earth is a long distance away. Wall Street's been jonesing for that mineral load for years."

Rob said, "What do you mean?"

"They want the raw materials." We must have looked bewildered. "They're free from an asteroid. They're expensive to mine on Earth and some sources are depleted."

"They're seriously working on that?" I said.

Lewis leaned back with his crooked smile. "I'll have to start teaching this, The Space Act of 2015." We laughed. "No, seriously. It gives the U.S. the right to private development of space resources. It's consistent with international treaty obligations. It's law."

"What's so valuable?" I said.

"Gold, silver, platinum, tungsten, iron..." (Lewis has eidetic memory, he can do this kind of thing.) "...cobalt, nickel, aluminum, titanium and enough hydrogen, ammonia, and oxygen to make rocket propellant. I'm going to make the class memorize all of that. Think of it, a shuttle service from Earth for raw material import. Huge profits."

"Yeah, well, we only need the asteroid to impact," I said. "They can have what's left of it."

So fine. We started planning.

"Directing an asteroid away from Earth is easy," Lewis said. "It's general. But directing one to hit a spot that spouts water is specific and basically impossible." Lewis puffed his cheeks then let out the air slowly. "It's really hard."

"What do they need from us exactly?"

"Coordinates."

"How do we get them?"

"Core samples and that means human hands. So I'm asking you two again, seriously. Road trip?

A little later parents were looking at me stunned. I thought it would be interesting to them.

"Just a little exploratory excursion," I said. "We'll be taking pictures we think you'll enjoy, areas you haven't seen..."

(Pinocchio's nose grows when he lies, doesn't it?)

They didn't seem interested. They didn't want more Mars, they wanted less. They wanted home. And anyway the daily trampoline had turned into their favorite distraction therapy. They were doing much better.

"You going to be safe out there?" somebody said.

"Of course. We wouldn't go if it was dangerous."

"Well if you don't know what you'll find, it could be."

"Kit means it's all familiar," Rob said. "Satellite feeds show us everything. We've already seen it, but we haven't been there. It's a big adventure. It'll be fun."

"It's going to be dangerous," my dad said looking at me, talking to me directly.

"My mother said, "Don't go, Kit, please."

That surprised me. She was an adventurer herself or she wouldn't have come here. And she's a champion. She isn't fearful. But she's my mother and not well and she feared for me.

I gave her a big smile. "Don't worry."

Under my breath, I said to Rob, "I didn't expect this. What do we do?"

Several other parents were saying we shouldn't go, dangerous, too risky.

"It's not risky," Rob said to them. "It's safe. Lewis said so and he knows. We'll be okay."

They went quiet. Some went on talking among themselves.

I said to Rob, "I wish Lewis were here. They'd believe him."

There was more quiet talking among the parents then Bethann's dad stood up and with umpire authority said, "We'll go."

Rob and I stared at him. Several other dads stood and said, "Yeah, we're the ones to do this."

Other men joined in, nodding in agreement.

"Me too."

"I'll go."

"Count me in."

It caught on--something interesting, challenging instead of dead-end lives in the infirmary.

A mother stood. "Count me in too," she said, bold and loud.

Now all the mothers stood, every one of them. Everybody in the infirmary was standing except for the few truly bedridden.

In the silence, I saw Mr. Elkins standing at the back door waving his stick.

"Me too. I can be your ears."

24

An hour later back at the lab I said, "You could see how good it made them feel, Lewis. They don't care if they ever come back. They just want to know what to take pictures of that we might want to see, what's scientific and what isn't. They've already started appointing duties, who'll be keeping records and who takes pictures and makes videos. They've decided somebody should record the entire trip, keep video going the whole time, day and night so they can send it back to NASA. That'd be a first, wouldn't it? Earth people forced to sit through Martian holiday pictures?"

Lewis listened, quiet, frowning.

"The deal is," Rob said. "All kids stay behind. They're glad to do something dangerous themselves, but they're not letting their kids do it, and that means us."

Lewis looked away.

I said, "What do we do?"

He continued to think a minute then he said, "We let them think they're going. That'll keep them busy while we prepare."

"And when we leave one day in a big rover with equipment...?"

"You'll leave at night and the next day it won't matter what they think. It'll be life or death for all of us which we still won't tell them."

That's when it hit me. This trip was serious. Up to then, I didn't feel it, residual trampoline euphoria maybe. Anyway, Lewis gave us his serious scientist look and made us focus.

"You're going to need a driller robot. I've got them working on that in engineering."

25

Because Mars is half the size of Earth, our mid-latitude location is 2,000 km (just over a thousand miles) from the pole. And we weren't going all the way to the pole, just toward it where Olympus Mons and deep, liquid water is. Olympus is a volcano 25 km high, that's sixteen miles. Earth's highest Himalayan peaks are less than five miles. We'd be able to see the top of Olympus soon after we left the colony.

What we were going to do had never been done and our route had never been traveled. But our satellites, the ones we monitor, and the Mars Reconnaissance Orbiter that NASA/JPL monitors, had explored it very well. So everybody in Houston and Pasadena knew more or less the location of the best possibility for deep springs. Why worry? There is water on our planet all right, a lot of it, deep under our feet. And a lot more of it closer to the polar cap.

Parents were making plans for their "fact-finding trip." It kept them occupied.

"Things look peaceful," I said standing with Rob and both doctors at the back of the big infirmary room.

Sylva nodded. "Most of them are looking forward to exploring."

Chang looked at us and said, "It makes all the difference..."

Noise interrupted.

Two couples were approaching Mr. Elkins who said "Who

goes there?" standing, stick raised.

Before we could move, all four rushed Elkins, seized him, tossed his stick and barged into the airlock. The door slid shut and locked for decompress.

We pushed through the crowding parents and Chang pounded on the door.

"Open up!   Right now!" as though his order would be followed.

A couple of parents pounded on the door too. Others drifted back and watched, wide-eyed.

The moment the lock released we entered, had to wait for decompress, then burst outside, but of course too late. Mr. and a Mrs. Shi and Mr. and Mrs. Eggers were lying dead, embracing.

***

Rob and I didn't eat that night. We slept badly. Next morning this latest multiple-suicide had to be reported to Jeff at JPL. Everything has to be reported.

"I wish there was a softer word than suicide," I said to Lewis.

"There is." (Good ole Lewis.)

> MARS: Good evening. Some bad news. Four more deaths, self-inflicted. Out the airlock door, no protection. You going to be able to keep this off your news?

JPL: News already has it. Anything personal is texted way ahead of you.

MARS: Even our water problem?

JPL: Not that. Your people don't know about it yet, do they?

MARS: No, and we don't want them to. They're having a hard enough time.

They talked about a couple of minor issues, then Lewis told Jeff our plan, the asteroid hit, the only thing we could come up with.

JPL: Glad you said it first. That's still all we've got..

MARS: Sounds like a B-movie, doesn't it?

Rob told Jeff we were excited about the trip and that we'd find the best coordinates for them to target. Jeff said they'd be aiming for the planet, about the best they could do. Then he pointed out that several asteroids would be in the closest position in a few weeks.

JPL: So get us some coordinates, okay?

26

Rob and I went to engineering every day to watch our equipment being built, parts printed, then assembled. We slowly began to feel better.

"You'll be in a large track," Lewis said to us. "It's a deck with a cabin at one side for drivers and six sets of track wheels. They rotate 360 degrees, travel any direction without turning the deck." He showed us the plan. "They go over bigger rocks than we have around here. Rocks are bigger up that way."

He clicked to another plan. "Lower these speed wheels and they lift you off the ground. Plenty of clearance for the undercarriage traveling on rocky ground."

"You're building a speeder and a grinder," I said. "Grinds over boulders, speeds over rocks."

"I guess."

Lewis was excited putting that vehicle together for us, custom built for our expedition. And beside speeding and rock hopping it was a driller.

"Coring drill," he said.

The drill was situated in the middle of the deck. All Rob and I had to do was position it over the optimal locations and remote control it. We'd be able to read what it was going through and when it reached frozen water we'd mark the coordinates.

"The depth doesn't matter too much, you know," Lewis said. "You'll just send the read-out to us and keep moving around the area."

"You'll be talking to JPL the whole time?" I said.

"Pretty much. When you've found the best spot I'll send the

coordinates. Then it's up to them."

Rob asked Lewis, "Can they really direct an asteroid to a specific spot?" No response. "I didn't think so."

"What makes you say that?"

"You didn't answer."

The engineering door suddenly swung open and it wasn't an engineer. It was Dr. Chang and a group of parents throwing back their hoodie-helmets and looking around. Lewis crossed the construction area toward them fast.

"Way too much work going on in here," he said.

"They want to see what you're building for them," Chang said giving Lewis a knowing eye about the secrecy.

Lewis gave in and nodded.

"All right," he said.

He gave them the same tour he'd just given Rob and me, showed them the tracks on the big rover how it could go anywhere fast or slow. And when somebody asked about the drill column in the center, he explained, making it as simple as possible.

"This bad boy can be used for core sampling but you won't need it for that. You'll just enjoy the ride. I thought you'd be more comfortable in a...sedan." With the side of his fist he gave the track a solid bump. "Very smooth ride."

Their eyes were shining. This was it. They were going to be useful again. It would take their minds off aging on a foreign planet, which is what they thought ailed them and to some extent it did.

Meanwhile, the trampoline was not entirely forgotten. Daily bouncing was obligatory. Chang told them it was to keep them in shape for the trip. For that same reason, they also continued working out in weight suits. They were loyal to their Earth exercise, lifting weights, doing push ups and chin ups, working on bikes and treadmills. To them, it felt like home, exactly like home.

That day walking around the track, admiring their custom built transport, one of them said, "We can't all get on there."

Here's how fast Lewis thinks.

"Quite a few of you can," he said. "And the others will be in small rovers, speedy ones leading the way. A scouting party preceding the big track."

That was even more appealing to them. Lewis suggested the whole fleet would take pictures and gather data to send JPL.

"You could make a tourists holiday guide for your friends on Earth," he said.

More eyes lit up. This was sounding better and better.

They finally left and when the last of them filed out, I said, "I feel shitty."

"You should feel justified," he said. "Come see what we're building for you and Rob."

27

A robot.

"Arbitrary drilling is useless," Lewis said. "You need a guide and that's Baby Bot. Plus she's a learner. She'll go out first and sample the area with a laser hand drill. It's powerful and reads out what it's going through. It can go deep but it can't give you samples."

"Sounds like a metal detector," Rob said.

"Kind of. Kit, you'll be in the cabin using these..." Interactive Robot sleeves were on a table near Baby Bot. "You'll control her arms from inside the cabin. Rob'll be outside with her."

I put on the sleeves. Lewis hit a couple of switches and the bot came alive.

"Move your arms," he said.

I did, and Baby moved hers. I kept moving, arms, hands, fingers--she had well-articulated fingers--and she did the same.

"Very old technology," Lewis said. "Very useful."

Baby had a slit for camera eyes in her shiny, egg-shaped, bronze head and instead of a lower body, she rested on a flat, box-shaped chassis with wheels. It contained the instruments to guide her movements along with mine.

"She'll do everything with the hand drill for you," said Lewis. "She'll use it to find the spot, then the coring drill will move in and do the actual sampling. At that point, you and Rob won't have to get out of the cabin. Coring's remote. And if Baby's not working she rides on the deck. The crane lifts her."

114

The crane was positioned behind the cabin and reached out over the deck. The track driller and the bot were built to do a job and keep us safe.

***

Daily, Baby Bot became more and more usable. When she was finished Lewis contacted me.

"Let's give her a try," he said.

I ran all the way to the lab.

Lewis connected Baby to a power source. She lit up, very real, very alive, able to do much more than the first time I tried her. Her arms began to move and so did her head. As her head moved, her helmet-slit camera eyes tracked everything she saw, which Rob and I could monitor inside the cabin.

I waved at the wall and so did Baby Bot. I waved at Lewis and so did she. He returned the wave. I turned her toward me and she waved at me on her own.

"How'd she do that, Lewis?"

He was grinning. "She's got chips that let her analyze her movements. I told you, she's a learner. Try going at her, attack her."

"I don't want to."

"Do it, Kit. She has to be able to recognize a threat."

I took off the sleeves and went toward Baby waving my arms,

thrashing at her. Baby backed away and held her arms crossed in front of her helmet-slit eyes.

"Good girl, Baby," said Lewis. "Exactly what she's supposed to do."

I stood looking at the bot, amazed. "She's my new best friend."

Lewis said, "Mine too.

Rob had just come in and saw the attack maneuver.

"What's this?"

"Look." I pulled on the sleeves.

I made Baby salute him. He returned the salute. Fun.

When no more commands were coming from me, Baby put her hands on her waist where it met the chassis loaded with her high-tech innards, her default position. Rob wanted to try the sleeves. He put them on and had as much fun as I did.

"Get to know her," Lewis said. "She's important."

It turned out to be the best day of Baby building.

That night I dreamed about the Cantina in Star Wars (one of our favorites, of course). Baby rolled in. The Cantina craziness went silent. Then, all the "scum and villainy" that Obi-Wan warned were there, attacked her. I woke up with a gulp, breathing hard.

Rob woke. "What's wrong?"

"A dream." Rob folded me in his arms, our default position.

***

The following days were productive--moving along fast and eating up survival time—they had to be. JPL contacted us and said they had an asteroid in their sights.

> JPL: It's the right size and in two weeks it will be the best time to get it moving. Your orbit will be near it and we need a target to aim at.
> MARS: Target? Getting precise, Jeff?
> JPL: Just want you to know, we'll be aiming mostly at the planet.
> MARS: Yeah, we got that.

There was no response and I thought they weren't sure they could do it or if they were, it might hit Rob and me out there somewhere. Rob must have had the same thought because he sent a message.

> MARS: Try not to hit the colony or us, okay?,
> JPL: Will do.

We signed off.

28

A couple of days later a crowd of parents pushed through the engineering door when I was waiting to practice with the sleeves.

117

Chang was with them and came toward me shaking his head.

"I couldn't keep them away. Sorry. Actually, they were out on the trampoline, having a good time, in a great mood these days looking forward to..."

"We got away, didn't we Doc?" A few parents had come over to us. "We snuck off."

Chang nodded and they laughed.

"We had to see what else is built for us. Don't make us leave, please."

Lewis didn't like this at all, a chattering invasion of his thoughtful, scientific world. The lab and the engineering building are his domain, we all know that, and until now no one had any interest in invading it. But nothing was the same anymore and Lewis nodded and let Chang lead them around.

Baby Bot was up on a pedestal being finalized, full diagnostics being run on her software. One of the engineers had the robot sleeves on and seeing the audience, made Baby give them a salute.

Their jaws dropped.

They swarmed her, a swarming pack with their explorer optimism. They wanted to touch Baby and talk to her.

Lewis turned to me and spread his arms in hopeless disbelief, all of this so out of line, inappropriate, undisciplined. But the engineers, mostly parents themselves who had escaped the affliction and were sworn to secrecy, understood. They managed the situation.

"She doesn't talk," one of them said.

They wanted to shake her hand.

"She doesn't shake hands."

That wasn't true. Baby had those well-articulated fingers, but the pack had to be kept at a distance.

Lewis went to them. "She's delicate."

"Can't be very delicate," somebody said. "If she's going out in the cold and radiation."

"Her metal is her protection, can't be damaged. That's why she's going along so you don't have to take those chances. She'll do what you tell her."

"I want one of those when we get back," said a voice.

"Me too," said several others.

Children in a toy store.

Seeing a bot built "just for our Mars exploration" made them even happier and more childish.

"Looks like the bot's ready. Are you?" I said to them.

"Yes, yes, we are," they said.

"Well, I'm not so sure about that."

I had special status with them since my parents' trampoline performance and they listened to me.

"I know for a fact that you'll need extra reserves of energy and strength out there day after day," I said. "Only prepacked food, probably less sleep..." They exchanged glances. "You'll be having fun but burning energy like crazy, I know. I've done research trips like you're going to do..." (ever more skilled at deceit) "...and you'd better get in shape. I can tell by looking at you, you aren't."

They exchanged looks again and laughed nervously.

"We're counting on you being ready for this," I said. "And the bot can't do everything. She can't energize you. And by the way, she was meant to be a surprise for you the day you left."

They went quiet.

"I want you to get back on that trampoline," I said. "I'm going to help Dr. Chang make up a work-sheet that we'll send to every one of you and you'll have to follow it and do the exercises every day. We're going to be checking your strength because you'll have to be fit. It's the only way you'll have fun and not get hurt, okay?"

There were scattered yeses and okays and a few humble looks.

"You can do this," I said. "JPL is ready to receive your reports and videos when you come back...(liar, liar, pants on fire)...so you'll have to work out like athletes or else Rob and I will go." They laughed at the threat, found it very funny, and kept saying, "Oh no, absolutely not," until Chang led them in a single, loud, obedient, "Okay!" and like dutiful children they left.

***

Next day I went to the infirmary and explained the check sheets I'd sent. I wasn't lying about those.

"On your pads you'll see a line for every day with a box for every exercise, diet and hours of sleep. Check off what you've done. No cheating."

A mom asked, "Are activities inside considered exercise?"

"Only if you're doing something you'd do at the gym or on the trampoline."

"In weights, right?"

"Of course."

Wearing weights would make them stronger, I reasoned, give them a shorter recovery time and more resilience. I realized suddenly they wouldn't need it, wouldn't use it. I'd been suckered in for a moment by my own lie.

As they filled out the forms, Dr. Sylva and I talked.

"What happens when they discover Rob and I have gone off with "their" track rover and "their" baby bot to do the exploring they've been looking forward to and exercising for?"

"Don't worry. You're doing the right thing," Sylva said. She was quiet and sensible. That calmed me a little. "It's the only thing we can do," she said.

"Until they discover they've been duped."

29

Three weeks later, Jeff seemed confident speaking to Lewis

with "good news".

JPL: Laser sublimation. Mirrors concentrate solar rays on a small area of the asteroid. The heat causes it to spew vapors and the thrust alters its path.

MARS: You got that off the Internet, Dude. Like a magnifying glass on an ant. The ant fries but our asteroid moves?
JPL: We put that on the Internet, Dude. We're working on a more sophisticated multiple mirror system. Should work well on an asteroid.
MARS: Should? When?
JPL: Soon as you give us coordinates for a target.
MARS: We're about to go search for that. We have a rough departure time. 48 hours from now. 0500. We have issues to deal with here first.
JPL: Don't leave until you iron out the issues.
MARS: They're not mechanical. It's the ailing parents and they're not an easy fix. Rob and Kit are set to leave in 48 hours and a mob of parents is going to be out of control when they do.
JPL: Roger that. Good luck. Five...ish is it?
MARS: Piss off. And get that rock pointed at us. We'll send coordinates.

They signed off.

We were quiet a minute then I asked, "What do you think it's like there now, Lewis? They worried about us?"

"Only JPL and NASA know so far. I don't think this will be out wide for a while. People love breaking news but they may not be interested in us. They've probably forgotten about us like we forgot about whoever was on the space station, or those last guys on the moon. It gets to be old. And they've got their own Earth problems, big ones. We're fifty people in danger. They've got hundreds, thousands, marooned in floods and hurricanes...two of those recently, did you know?"

"No, I didn't."

He shook his head and took a breath. "Okay," he said. "Day after tomorrow, five...ish." He wasn't smiling.

***

Rob and I sat together in my hab that evening.

The face of a U.S. governor was on the screen saying, "I've declared a state of emergency and our Coast Guard is on alert. Some lives have already been saved..."

It switched to video of people on their roof or in small boats and an announcer standing in front of a U.S. map. She indicated the areas along the West Coast, East Coast and the entire Gulf Coast.

"They're under water in all these locations," she said. "And

123

more heavy weather is coming."

A simulation of a hurricane approaching the Bahamas appeared, it's path arcing toward Florida.

Rob switched it off, "Glad we're here."

***

In the two days before we left, we heard often from JPL. They had suggestions and advice but nothing useful or that we didn't already know. They asked more questions as they continued to work on the asteroid redirect. They were paying plenty of attention to us even if Earth's general population couldn't. NASA is loyal to its people, especially the ones it hurtles into space and we know that. We've never lost sight of that.

"It's keeping them awake at night," said Lewis.

The afternoon before leaving, NASA contacted us saying they were sending something. Lewis said they were worried.

"Guilty?" Rob said.

"Maybe. They should be, fucking faulty-cured concrete...had to be something faulty in the presupplies. They're sending a small probe now with...something."

"Why do that?" I said. "We won't have it for months. What is it?"

"I don't know. They feel bad and want us to know help is on the way, I guess. A million dollar get well card."

30

Rob and I were definitely not going to mention any of that when we went to visit our parents that evening. Mine wanted to celebrate my birthday.

"Almost eighteen," Mom said the day before.

Dad had smiled and said, "My little girl, all grown up."

I understand eighteen is something special to them, an Earth thing, but I was much more excited about our departure at five the next morning.

Before seeing them we had one last practice session rehearsing everything again, over and over; track, drill, robot sleeves for Baby, emergency procedures for everything else. We needed to be checked out on all of them. Again. In secret.

"Redundancy is good," Lewis reminded us then repeated, "Baby's entirely remote. You can sit in the cabin and toggle her. She'll be out there handling the drill while you monitor data inside. Simple. When it seems safe you should be out there with her, Rob."

We listened carefully. It was established that I would use the sleeves and Rob was for outside procedures, all this going on hidden behind the engineering module. Baby was locked down on the deck. Rob and I were inside the track cabin in full gear.

"Lower her," Lewis said.

I flipped switches and the track crane emerged. Its big tow

hook lifted Baby off the deck by the top of her bronze helmet head, swung her out, and lowered her to the ground in front of us.

"Put her back, Rob."

We both needed to be skilled at handling her. We'd practiced many times but Rob flipped the switches again and this time as the track crane lifted her toward the deck an alarm went off. Rob and I jumped toward the door.

"The cabin's an airlock," Lewis said. "Is it on decompress?"

Rob stopped himself and checked the control panel.

"Yes."

"Who set it?"

Rob looked at me. I shrugged.

"I don't know."

"You'll have to get used to this, both of you. During the day when you're working and driving, you'll be wearing skins and glass all the time. The cabin will be in decompress so you can jump in and out when you want. At night you stay in and pressurize and take off your gear to sleep. Next morning you'll suit up and lock the cabin on decompress for the day. Got it? It's like you're always outside during the day, even when you're inside."

"We won't forget."

"You might, so there's a timer and a warning light. After you lock yourselves in at night you'll have to release the timer to decompress and go out. You have to remember this, both of you."

"You don't trust us."

"I don't want you to blow yourselves out of the cabin."

The rest of the session we practiced lifting Baby off the deck and operating her on the ground. There were more surprises from Lewis with his remote and we learned from them: be ready all the time for the unexpected.

"We're not astronauts, Lewis," I said at one point.

"Yeah you are. This is entirely new, different from living in the colony. Make a mistake and you die pretty much like astronauts."

Drill instruction went on for another hour then Lewis said, "Okay. Fine. Handle everything up there just like we do here. Weather's good, just some mild dust activity toward the cap. You're not going that far but you'll be in frost the second day. That's as far as you go. It's colder but you're ready for that. So is Baby."

31

A little later: "Happy birthday, honey."

"Thanks, Mom." She was sitting on her infirmary bed alert and reasonably happy.

Rob went to his mother, still sitting unmoving, staring straight ahead.

Dad said to me, "How is she?" and I shook my head. "Tell Rob to join us."

"He's going to."

"Good. So, Sport, do you feel grown up?"

"I do, yeah."

I felt no more grown up than ever, although I'd certainly learned to lie. Is that particularly adult? I think when Rob and I first got together, that was grown up but we were so young. On Earth it's a big thing, I know, making love the first time, but here, inoculated, we experiment with sex really young. It isn't taboo like on Earth, maybe taboo isn't the word. It's natural here because it causes no harm, no disease, no pregnancy. When I learned about those problems on Earth, I understood the taboo. Now, of course, you have inoculation there too, but I wonder if the attitude has changed.

"I do feel grown up," I said, not lying.

Going on an expedition to locate water, find where it is and how deep; that's grown up.

Driving a track rover and running the arms of Baby Bot that my parents, all parents, had been practicing with and having such fun with; that's grown up.

Knowing I'd be making off with all that equipment in a few hours felt very grown up, but the lying felt sour and wrong and killed any pride I had in doing it.

"We have gifts for you," Mom said.

That snapped me out of my dark thoughts.

Rob joined us, and I opened a present, an album Mom and

Dad put together about their life before Mars which they knew I couldn't quite comprehend or admire.

"You were so brave living there," I said.

"We thought coming to Mars was brave."

Among the photographs in the album was a note: Look under the bed for another gift. I did and found a virtual reality device for two people.

"I've never played around with one of these," I said, honest enthusiasm.

For parents, VR is the only way to go home again which they'd often done. But for me, VR was a game, a new experience.

***

Rob and I were walking on a beach "somewhere on the French Riviera" holding hands. We could virtually feel the sun on our faces. We could virtually see the water.

"This is wonderful."

Hard for us to imagine a large body of water but we were walking beside one, a VR Mediterranean that seemed very real. And the VR sand and beach were so like our red Mars sand it soon seemed not only real but nice. For the first time we were on Earth with no danger, no speeding vehicles or threatening wildlife. My parents had given us a beautiful, safe, Earth environment.

Still wearing headsets, we told them how much we liked it, "No wonder you miss it."

Rob and I strolled the Riviera beach passing virtual couples, no one wearing gear, of course, sun on our faces, air fresh. Then all this was interrupted. A cake was put on the bedside table and the whole infirmary sang Happy Birthday. They clapped and I thanked them.

"Could I go for another virtual walk before I have cake?" I asked.

In minutes I was in a wooded area, birds singing around me, wonderful. *Of course, they missed that.* I understood.

Rob had been captured by my dad who had him watching a movie on the wall, one of his favorites. A man with a rifle was aiming through a scope that he was calibrating.

"He can hone in on a speck a mile away," Dad said. "And if he's a good sniper--well, they're all good--he takes a breath, almost stops his heart, then squeezes off the shot. Amazing, those guys."

My super-athlete dad admired any kind of excellence, certainly marksmanship.

"What were guns used for exactly?" Rob said, the big question that all Gen2s asked, conflict being such a big thing on Earth.

"Killing the enemy, shooting a rabbit, target practice for fun. For crime, of course," shaking his head. (My scientist Dad did none of those things.)

Rob nodded.

"So war kills off people so that different people can live there?" Rob said.

"Well, sometimes, but it was usually political."

That was how war talk usually ended, no resolution at all, still a mystery to Rob and me. To all Gen2s.

Rob and I went walking on the VR beach again, no war.

32

Later that night, back in my hab, we sat together, quiet.

"Seven hours," Rob said. His face was glazed.

"You okay?" I said.

"Yeah, sure. Why?"

"You don't look okay."

"He got up and looked in the mirror on the kitchen wall, the one I made when I was little with flowers as I imagined them, painted across the bottom. Rob's frowning, slightly shiny face appeared over the flowers and he laughed.

"Yeah, I look worried. Guess I am."

He splashed water on his face, dried it and ran his hands through his hair. He plopped down on a chair, elbows on his knees, looking at me.

"It's risky, you know," he said.

"I know."

With a contact ping, Jeff was on my wall screen.

JPL: Anybody home?
MARS: We're here, Jeff. Ready to go.

We told him we knew making the asteroid hit a target or hit Mars at all was like throwing a dart out a window; random, might hit anything or nothing. But Jeff was upbeat. He wasn't buying the uncertainty.

JPL: Hey, it's going to work. We're all set here. Mirrors in position. Asteroid perigee in a few hours. Everything is waiting for your coordinates.
MARS: You sure about that? It sounds like getting any hit at all will be difficult.
JPL: Everything we do is difficult but we got your parents there didn't we? We need target coordinates, the best you can give us. We'll shake something loose.

'Shake something loose' sounded confident to me, even relaxed. (As I write this now it seems completely arbitrary and totally dangerous.)

JPL: Remember, we'll be on your heads-up display all the time, every minute. We'll be with you watching and listening to everything you say, everything you need, any worries you have, anything you run into that's trouble or that you need to know. So watch your display the whole time. You'll see us with you. You're not alone.

"They'll be fifteen minutes behind," Rob said. "What's he talking about?"

"Just trying to make us feel better."

"Or it has leaked out and we're big news and that speech he just made went out to a big audience."

We switched to a wall image and ran Earth news. Yes. There we were, a picture of the two of us standing together in our skins and hoodie-helmets. Lewis took that picture, proud of the new gear.

The voice of an announcer explained.

"They're searching for flowing water and NASA is sending an asteroid to shake it loose. These Martians are crafty and sturdy. They've been there twenty years and they're flourishing. Now they have a problem and their kids are going to help fix it. Odds are, with NASA's help, they will find their water."

The camera panned down to action figures of Rob and me in skins and face glass. We were stunned speechless.

"These two Martian teens are the new superheroes. Since

we've heard their story, their silver gear has become the favorite party costume complete with a smoke ejecting jetpack. And Disney has announced it's adding a dust surfing attraction to their Orlando park. Competitions expected."

We doubled over laughing and amazed and didn't hear another word until we noticed a map on the screen beside the announcer and his serious face.

"...the problem in South Florida, New Orleans, and lower Manhattan, continues to be too much water. The current conference in Brussels with EPA and Dutch experts is expected to result in new flood walls in place of failed levees and dikes. They're to be built as soon as possible although the Army Corps of Engineers are stretched to the breaking point and that makes for..."

Rob switched it off.

"Thanks," I said.

"Their problems are bigger than ours." He shook his head and smiled, "Action figures."

"We're comic relief."

33

Departure 0500.

We were ready, suited up, standing with Lewis at engineering. The track was rolled out, ready.

"Bigger and badder than ever," said Lewis.

Baby Bot was locked on and supplies were stowed. Two mechanics were running diagnostics and checklists--it felt good seeing that. I kept thinking they'd be the ones held responsible if anything went wrong. Or Lewis. Or the team at JPL who were watching in Pasadena; evening there. NASA's lifesaving laser sublimation was going to put to the test, not to mention their future funding...

"Kit!" Rob snapped me out of my thoughts.

"Yeah?"

"Listen to Lewis." I was drifting. Not good. Terrible for what we were about to do.

"I'll be watching," Lewis said. "Anything you want to know, ask me, talk to me. Talk to me all the time. Watch your HUD more than where you're going. The track knows what's in front of it."

"JPL said to watch them," I said.

"Why? They're fifteen minutes behind. Watch me."

"I was kidding."

"Don't kid." He gave me a hard look. "Reposition the head-up display on your face glass so it's almost in your line of sight."

We did and Lewis nodded. Two engineers stood watching. We climbed in the cabin and pulled down the track dome. It locked. Through it we saw the last humans we'd see for a while. Forever maybe.

Lewis gave me a thumbs up and an A-okay nod.

I'd done this in practice. This time for real, I powered on and the instrument panel came alive. At Lewis's nod, I moved the big track slowly forward. Rob was checking the programmed route that JPL thought would most likely lead to underground water. He wasn't looking outside. I was, and I waved at the parent engineers. They smiled and waved back as we moved off, then they gave me a thumbs up. I realized too late there was no waving when astronauts blasted off from Cape Canaveral.

What a novice I was at all this.

But we weren't blasting off and we weren't astronauts. My wave embarrassed me but those dads knew I wasn't familiar with space travel and it wasn't space we were going into although it was almost as unknown.

I wasn't concentrating. Rob was.

"Kit," he said and tapped the time readout so I'd notice. 0500 exactly. "Take that, NASA." He grinned.

That made me feel better. Okay. So this was going to be fun.

34

Back at the infirmary parents were asleep. Rob contacted Chris Chang who was awake waiting to hear from us.

"How are you?" he said.

"On our way," Rob said. Did you see?"

"Yeah, I watched. You're gonna be all right."

"Everybody keeps saying that, but nobody can possibly know."

"We think it's true."

I put my head close to Rob. "Hi."

"Hi, Kit."

"This is kind of like talking to my granny on Skype without the delay," I said.

"Let's hope," said Chang.

"Just hope? You said we were going to be all right."

Chang smiled and I went back to driving.

Rob said to Chang, "What's your plan for when they wake up?"

"Let me see if it works first."

***

The speed wheels were lowered, racing us toward the pole. They bounced over rocks and threw us around. If we weren't belted in we'd have hit our head on the top of the cabin a few times. It did happen to Rob before he adjusted his straps. His neck was sore for half an hour. First error.

"Lewis should have warned us," I said.

"He told us to always buckle up."

We drove, bounced, held on and hardly spoke. It was uncomfortable, not dangerous. We reached the latitude where Lewis said the tracks should be used. We stopped and checked in.

"Can we elevate the damn speed wheels?"

"Yeah. Do it. Was something wrong with them?" He was looking at a digital elevation map. He saw what we were driving over.

"No, they're fine," Rob said. "Just heavy bouncing. We've got bigger rocks now, like you said."

We raised the speed wheels which lowered us onto the track wheels, a slower but smoother ride. The big tracks independently crawled over boulders, rocking us but keeping us mostly level and comfy.

We came to the dry river beds. There are shallow ones outside our colony but these were deeper. And they all once carried water.

"Look at that," Rob said.

"Yeah. We're gonna find water."

Baby was riding in front of us, locked down on the deck squat and flat on her chassis, her wheels retracted.

"Our helper," I said.

I'd fully anthropomorphized her. Lewis said it was okay, would make her more useful.

I drove into one of the riverbeds to get a better idea about it, smooth, cracked, hard. Fifteen minutes later JPL saw that and loved it.

JPL: Stay in there for a while. We've never seen riverbeds from that perspective.

They'd only seen what the landers of NASA's Mars Exploration Program sent back over the years. Now, no matter what happened to us, they had this new record to study.

And they were right about us not feeling alone. I didn't. Lewis was there, real time, of course, and NASA/JPL was there too, but trailing behind. We were definitely not alone.

"Watch out," Rob said.

A waist-high boulder was ahead of us.

"I want to go over it. The tracks are made to go over it. Okay, Lewis?"

"Yeah, go."

The right front track rocked up and over, only slightly tipping us, and apart from the dust it rose, nothing was changed.

"I liked that," I said.

"That was a volcanic, igneous chunk from the core. You can go over any of them."

"Easy for you to coach from a distance," Rob said, smiling, liking this as much as I was.

A few smooth dunes appeared in the distance, barchans, they're called.

"Should've brought our boards," I said quietly.

Lewis said, "I heard that."

***

By afternoon we were out of the river beds and on the open plain again, fewer but bigger boulders. I drove between them.

"Speed wheels?" I said.

"Yes," Rob and Lewis said together.

We made double time toward the polar cap snow.

For all of you on Earth, the snow here is frozen carbon dioxide powder mixed with water ice, very slippery and not much of it.

We stopped. Didn't get out of the cabin. We munched food bars, sucked on water containers and felt sleepy, but we took caffeine pills and pushed on. Lewis never signed off for more than a few minutes.

"Derek's teaching," he told us. "We don't want the kids watching...in case..."

"I case we die? Ah-ha! So it is bullshit hokum what you guys have been feeding us about being all right."

"Use the word bullshit or the word hokum," he said. "Don't use them together. Your Earth jargon isn't very solid, Kit."

"That's a cover-up for you stay-at-home wimps. Oh, can I use all those words together? We're out here and you aren't. That goes for you NASA nerds and Jeff too fifteen minutes from now."

"You're pretty high," Lewis said.

"Caffeine. I'm not used to it."

"Then stop, don't drive anymore. Seriously, Kit, get out and move around."

35

We were still in skins, as trained, in case we had to hop out for some emergency. We released the dome and climbed out, stiff. We did some stretching and running in place.

"Take out Baby," said Lewis.

"It's almost dark," I said.

"Take her out and work with her. You might need to work in the dark."

"We practiced that."

"Not in this temperature and terrain. Get her out, Rob. Put on the sleeves, Kit."

I got back in my seat and we went through a session with Baby, Rob outside with her. I made her give him a wave and he waved back. I switched her receiver to my HUD so she could see Lewis waving to her and she waved to him. She could see JPL for the moment and I wondered if they'd wave back fifteen minutes later. Baby computed all this with her eye-slit cameras and waved again on her own, long before NASA could see.

"She can override my movements for something like that?" I said to Lewis.

"Yeah. She's learning. Now if she handles the drill just as well..."

"Should we give her a try?"

"Sure. Send her off someplace."

Rob hooked Baby's hands around the drill handle. They auto-locked and she was ready. Telemetry in her chassis rolled her away from us. The sun set and it was sudden darkness.

"Put some light on her," Lewis said.

We powered the track's spotlights and it was like day in front of us.

Arms in the sleeves, I sent Baby to an arbitrary spot, positioned the laser drill upright and like a gyroscope, Baby kept it perpendicular while it penetrated the permafrost and drilled straight down. Rob stood away from her but watched closely. I read the drill's data coming into the track. Lewis read it too.

"Happy, Lewis?" I said.

He had to be elated, his robot and his track and his students all performing perfectly.

"Well done," I said to him when he didn't answer. I took my arms out of the sleeves leaving Baby on her own.

Entirely focused on the data coming from Baby, Lewis said, "Nothing to indicate water below but we didn't expect it where you are. Good work."

Since Lewis doesn't do compliments, taking or giving, that "good work" meant a lot. Rob came back in the cabin with me and answered some questions Lewis asked about Baby. When I looked up she was gone.

"Where is she?"

There was no sign of her. Rob jumped out. I put my arms in the sleeves.

"Turn the lights on all the way around," Rob said.

I powered full lighting and there she was, behind the track.

"Why did she go there?" I said, rattled. Baby was the one thing I was supposed to control.

Lewis said, "Don't panic. I did it. I made her wander."

"Dammit, Lewis," I said. "Why'd you do that? I don't trust myself anymore. I blew it."

"You didn't blow it. You didn't do it. I built in a couple of tests for your first day. She'll roam if you don't keep track of her. And I made her roam, you didn't let it happen."

Rob was climbing back in and said, "That's not cool, Lewis. You scared the shit out of both of us."

"You let her out of your sight. Just because she can learn, don't think she knows what she's doing. She's like a child, a toddler. It makes her more effective. You need her to stay that way."

"Well adjust that, will you? Grow her up a little or something."

"And you did make a mistake. Know what it was?" We didn't answer, didn't know. "You're smart kids with no experience doing

143

this kind of thing. You took your eyes off the event. The event was Baby. You can't do that out there, ever."

"Too late," Rob said.

"No, not too late. You're learning. There's no harm done this time so learn from it. You'll need her built-in roving eye when you get farther north."

Baby was happily rolling back toward us. (You can delete that "happily" if you want, NASA. It's just how I felt about good-natured Baby by then.)

***

After a night of deep sleep on pull downs, our memory foam sleep-shelves, and morning caffeine (most widely used drug in the solar system), we were alert and discovered we could see Olympus Mons.

Wow.

The sudden sundown darkness of the evening before had cloaked it but in the morning's sudden sunlight, bang, there it was, even more amazing to you on Earth, I'm guessing. Let me remind you: sixteen miles high and the size of Arizona. Those metrics should mean something given your tallest Himalayan peak is less than five miles. Even for us, it's staggering.

"First humans to lay eyes on it," Lewis said. "Congratulations."

Fifteen minutes later we heard applause from JPL. Rob and I looked at each other.

"We *are* heroes," he said.

Gen2s are the first extraterrestrials but we're well aware it's nothing heroic, nothing we've earned. Neither is seeing Olympus. It was Lewis and NASA/JPL scientists and engineers and inventors everywhere who put our parents here and the only entitlement we have is our parents' coupling. Laying eyes on Olympus that morning did make us feel useful, though, just not superior. Gen2s only need to feel useful. It's much more pleasant than superior.

(Maybe not on Earth.)

It was obvious the team at NASA and JPL were focused on us and maybe all of Earth too if they knew what was happening. We thought even in the flooded areas they might be watching and witnessing the first look at Olympus and feeling proud. I hope so because your Earth tax dollars put us here.

Back at the infirmary, feelings were not so upbeat.

36

"I have an announcement, everybody."

Dr. Sylva wanted to tell them about what had happened after they finished breakfast thinking bad news on a full stomach and a good night's sleep would be easier to take. They had to be told before they discovered their track and Baby Bot missing.

Silva got their attention and said, "There's been an unusual development."

She didn't quite know how to begin and some joker said, "The trampoline is stolen?"

Everybody laughed, Chang too, playing along with them, trying to keep it light.

"No, no. Nothing that terrible," she said and went serious.

The parents went still, waiting, sensing something not quite right.

"They've left," she said.

"Who?"

"Rob and Kit."

It was my dad who said, "Left for where?"

"I don't know exactly," said Sylva. "They left a note for us and said they thought what you were going to do would be too dangerous and they'd do it instead. They can explore faster and safer than you can and they're going to bring back plenty of..."

Dad interrupted, "Show me the note."

Sylva and Chang were ready for that and handed him a note addressed to both doctors. Dad read it as Sylva told the others what it said.

"They're worried about you. They don't want you to get hurt. They've been more affected by your illness than you know and they don't want anything hurting you again or making you unhappy." There was muttering and talking. "They left early.

We didn't know anything about it until this morning."

She spread it on pretty thick telling them how more than anything, we wanted them to be well again (that was true) and knew they weren't yet (also true) and couldn't let them take the risk of this adventure (absolutely true).

"They promise to bring back beautiful pictures and videos," she said. "And as one of you suggested, they will keep the video on day and night. They like that idea." (Cameras in our skins and glass and all over the track were automatically recording everything, never a matter of choice.)

Loud talking broke out, objections to the news, even tears and anger.

"Are you saying they took the track?" said a mother.

"Yes."

"And Baby Bot?"

"Yes."

That was too much. They too had anthropomorphized Baby. The bot and the track were their property or felt like it. They'd practiced with them and learned how to use both them like Rob and I did.

They crowded around and wanted to know more, wanted to read the note which Sylva willingly gave them. They passed it around and felt duped. Many had tears in their eyes. They were betrayed and the note's benevolent motivation meant nothing.

"They were worried about you," Chang said to us. "Now they're just angry. Very angry."

Rob said, "So what do they think?"

"That you snuck off with their track and Baby Bot, and your reasons aren't very good. They're pissed off."

"Yeah, well, what you and Sylva told them was good. I don't know what else you could have said. And if we get back and this works we'll be forgiven. If not we'll die, so."

"You going to tell them what we're really doing?" I said.

"Not while you're doing it. They'll be worried on top of angry. Remember what all of us went through before we coming here."

"What do you mean?"

"All the settlers, rehearsing in a biosphere environment. No smoking, no alcohol, no meat for 90 days to see if we could take it. They were making us into a bunch of teetotaling vegetarians and none of us started that way, one or two maybe, but that's what we'd have to be and they had to know who could handle it. Those who couldn't washed out."

"How many washed out?" I said. "I never heard about this."

"Three. Everybody else just bit a bullet except for a few who were already super healthy, easy for them. My point is, now that they're confronted with a situation, they're furious and all that bullet biting they built up is breaking down. I heard couples in arguments they had when they first married. Old stuff is coming up. It feels like they'll cut loose any minute and riot again. I've got Dolf and my three strongest medics at the door. We're on alert. Mr. Elkins is helping me...he knows everything."

"I won't say a word," Elkins said, his head at the edge of the screen.

Rob said, "You and Sylva did good, Chris. You too, Mr. Elkins." Then he sent a message to JPL.

***

MARS: Are you ready for us?

JPL: Got two big rocks picked out. Maneuvered one already. Mirrors are ready to point and shoot at the other one.

MARS: Have you successfully done this before? You didn't tell us.

JPL: Several times. We've redirected big ones and they veer off.

MARS: In the direction you want?

JPL: Pretty much. Away from us. Laser sublimation is not sniper accurate.

MARS: So 'pretty much' is your accuracy factor?

JPL: It should work. It won't have to travel far. Give us coordinates. That's what we need.

MARS: That's what we're working on. Baby goes out for more drilling today. We're close to underground springs, we know that.

149

JPL: Right. Sats confirm it.

MARS: Go back to work, will you? Work on 'pretty much.'

Rob signed off. So did Chang who had a rebellion to deal with.

37

Baby was moving out across the frozen Martian regolith into a thin layer of snow and we followed, smooth ground, easier to negotiate than the dry river beds.

We could see what Baby saw with her eye-slit imagers. Lewis and I and eventually JPL read the information coming in from all her sensors. I was wearing sleeves, controlling her.

"Let her run herself," Lewis said to me. "She's programmed. She knows what she's looking for."

Rob was out there with her, watching. She was moving her bronze head shaped like a centurion's helmet, pointing her imager-slits toward the snow patches and then the surrounding areas just like a human would do looking for telltale signs but with no interfering thoughts about what's happening at the infirmary.

"She's amazing," I said.

She stopped moving. Her readouts which we could all see told her she was at a promising location. She positioned the drill and started.

I was twenty meters away. Rob was near her. The sound was faint of course in the thin Martian air. This was exciting, Baby's first find. The drill went in slowly through the permafrost, then a little faster, then faster still. A thread of dark dust spurted up around the drill. At first I thought it was a natural ejection from the drill, then I heard Lewis's urgent voice.

"Surface geyser."

A strand of melting frozen carbon dioxide turning to gas was sending a string of dark sand straight up. It instantly became thicker and rose above Baby's head.

Rob shouted, "Get her out of here, Kit!"

My robot sleeves took her over.

Lewis ordered, "You get out, Rob."

Rob ran back as the stream of dust became more harsh and powerful. It hit Baby's helmet blocking one camera.

Lewis yelled to me, "Kit! Get her away!"

But with one camera out she'd gone into safe mode. I couldn't make her do a thing.

"Get her away," Lewis yelled again.

"She's safed," I said.

"Go get her, Rob."

The geyser was growing in size and strength, no longer a thread, but an arm-thick, jet that had knocked Baby and her chassis back flat, still frozen in safe mode. Nothing I did could move her and nothing about her was safe.

"Rob, it'll get bigger," Lewis said, controlled but intense.

Rob was with Baby who was lying back on the edge of her square chassis, staring at the sky. The geyser gave out a thump, doubled in size, and knocked Rob away. More thick sheets of black dust showered down.

"Drag her back, Rob," Lewis said. "Pull her back."

We could hear Rob's labored breathing as he ducked into the downfall of dust and grabbed Baby as he would a human under the metal shoulders and pulled. She didn't budge. The drag of the chassis assembly was too much. He tried to lift the chassis upright so he could roll her away but it wouldn't move.

"...too heavy," he said.

I left my seat to go help, and Lewis stopped me. "Stay at the control panel, Kit. She's stuck on something, Rob."

"There's nothing to...what's she stuck on?"

"I can't tell."

I hit switches and moved the track in close. The black sand showered down across the dome. I lowered the crane. It didn't reach as low as the chassis, missing by a meter. I set a timer-- everything on the track can be used remotely--and I hopped out ignoring Lewis calling after me.

"The drill," I said to Rob.

We hooked the T-handle of the drill under the edge of the chassis, and the two of us held it there as best we could. The chassis's edge had a very small lip.

The timer started the crane and it raised Baby's chassis. The higher it got the less acute the angle holding the T-handle. It looked like we were at the tipping point when the hook flew off and hit Rob in the chin. He fell back. Baby's chassis balanced a second then fell level, wheels on the ground. Rob was dazed, almost unconscious.

As I dragged him away from the curtain of falling sand, his head cleared. I went on in the cabin, hooked Baby by the helmet and lifted her away from the spewing geyser onto the deck. Then I helped still dazed Rob into the cabin and backed the track away from the geyser. We sat trying to breathe, covered in black dust and sand. Rob's chin was bleeding. The geyser continued to send up a curtain of dark material 30 meters high.

We heard Lewis calling us.

"Answer me. Talk to me! Are you okay?"

"Yeah, yes...I got clocked, that's all," Rob said.

"Okay, Jesus..." he was quiet for a minute. "That was bad," he said. "Really bad." Another minute. "Good work, Kit. I don't know if I'd have thought of that."

"Think I ruined the drill?"

He laughed, Lewis actually laughed.

"Jesus, no. Don't worry about the drill." He was shaking his head. "You're both okay and...that's good." His voice cracked.

"Baby's okay too," I said.

"And Baby too."

153

Lewis took more time, got himself together, then said, "Okay, you two look terrible. Clean up and take care of the chin, Rob. Kit, move the track farther away."

I did.

"Now listen," he said. "I want you guys to take your time and settle down. Eat and go to sleep for the night. We'll check out Baby in the morning. You need rest before we go on with anything."

We nodded again, still pretty dazed.

"Good work," Lewis said. "Really good work..." He wanted to say more and finally came up with, "Luck and good thinking...mostly good thinking...that's what...yeah..."

"Okay," Rob said.

There was an empty silence then Lewis said, "Oh, yeah. Set your audio volume low for a while."

"Why?"

"You should sleep, and JPL and NASA when they watch this..."

"Right...right."

38

It was Lewis's voice that woke me the next morning: "We can fix this in a day."

He and Rob were happily talking about what had to be done. Rob's jaw was bandaged but he was rested and so was I.

"Do we have an extra day?" Rob said to Lewis.

"We're not on a schedule and we have no choice. It'll take a day to fix everything and run checks. Looks like it's only disconnects, nothing missing or destroyed."

Seeing everything Rob's camera saw, Lewis went to work using Rob's hands as his hands on the tools, and it did take all day. During a break, Rob contacted the infirmary.

"The parents watched," Chang said. "We nearly died here."

"*You* nearly died?"

"Yeah, all of us watching."

"Why'd you allow that?"

"It was the right thing. They don't know exactly what you're looking for and there's no sound, so they can't hear what you're saying. I told them it was something about NASA security and we're lucky we can watch, so they only saw what happened and how you fixed it and it was perfect, believe me."

"No, it wasn't," I said. "It could have been fatal."

"I mean it was perfect for *them*. They think you saved them from what happened to you. When that geyser spouted nobody moved. Then when Baby was finally upright, they cheered, and somebody said wait till NASA sees this..."

As Chang went on it was clear that seeing what happened was good for the parents.

"They were in tears," he said. "Very moved. They held their breath, then when you fixed things they whooped and hugged each other like, I don't know, something out of one of their movies. But it was real and everybody was safe including all of them so...you made it right. They know they'd be dead if it had been them trying to handle all that."

"That's amazing," Rob said. "Glad to hear it."

"You've got a cheering squad back here so don't worry about them anymore. Kit, you're dad is puffed up proud of you, so's your mom. Rob, your mother...I only wish she could have seen it."

"No change, huh?"

"Nope."

"Maybe you should replay it for her. It might wake her."

"I'll wait till you're back. If she wakes up now and knows the danger you're in..."

"Yeah, you're right."

We signed off and contacted JPL letting Jeff know Baby was almost repaired. Rob and Lewis finished her and Baby was once more cleaned, reconnected and locked down on the track's deck. We ate and slept.

***

I had a dream. It was about my parents when they were still happy and not yet affected by the low oxygen supply in the grower dome. Almost every evening they'd sit close together comfy on a hab sofa. One evening, I was about six I guess and sprawled on the floor nearby, chin in my hands watching a wall screen movie with them. It was an old one about the West, the Old West they called it "with cowboys and Indians."

"This one is good," Dad said. "Watch it close, Kitty."

I could see it was different. The man befriended a wolf and even some of the Indians. That was different from other westerns. And his woman was an Indian woman, that was different. With all these various kinds of people wearing assorted kinds of clothing and all kinds of animals roaming around, I didn't like it. Too complex. Confusing. But I watched to please my parents.

The men rode fast on horses alongside a huge number of buffalo and shot some of them.

"For food and their skins," Dad said. "For survival."

Those words were the most vivid part of the dream. Then I woke up.

Rob was asleep above me as I lay there wondering why I had that dream. Because we were hunting, maybe? But we were hunting for water, not food and skins. Then I realized. Of course. What we were doing was for survival too, like the hunt in the dream. But nothing had to die in our hunt and I went back to sleep.

It must have been a deep sleep because the next thing I heard was Lewis's voice again in the early morning.

39

"Wake up, you two. Good news." Lewis was barking at us from the cabin's head-up display.

"What news?" Rob said, groggy.

He slid down from his sleeper.

"NASA thinks you'll get an H2O signature on D-G."

"Dodo-Goldilocks? Phoenix got a negative there decades ago."

"They've gone over everything again including Phoenix and they're saying you definitely should give D-G a try."

"Skip Wonderland? You sure? Wonderland's where we're headed. That was the plan."

"Only vapor deposits, there."

"We know that but we think it's worth a try."

"Go straight to D-G, Rob."

"You sound very certain."

"NASA orders."

Rob said, "Did you tell them we have a little wet lab on board and an optical microscope? It's pretty basic but we can analyze..."

"No. Watch this."

Lewis replayed his earlier communication with JPL for us, minus the delays.

LEWIS ON MARS: I don't like the idea. Phoenix showed there's perchlorate but no deep water. What's the point?

JEFF AT JPL: They've got the laser drill and it'll tell them more. We've reviewed the telemetry and found indications we missed.

LEWIS: Give me a minute, okay?

While Lewis checked his own data, Rob said to me, "Reviewed the telemetry means they don't know."

"Don't know what?"

"Whether there's water on D-G or not."

Jeff got tired of waiting for Lewis.

JEFF: Look. We've got quite a bit going on for you. The asteroid redirect is on track but we've never done something as precise as this. We think Rob and Kit have done great so far, really great how they handled the mishap last evening. And now that Baby is up and running they need to go to D-G fast as they can. It's not the best spot, we know that, but we've just found something in the telemetry review indicating that at the right depth, an asteroid will unearth a spring. It's possible.

LEWIS: Is it probable?

JEFF: Seems to be. Right now the best thing we can do is put an asteroid on a bearing for D-G. We've got data only on Wonderland's regolith. But the ice on D-G looks better. So tell them to go there and see what you can detect with the drill. Right away.

LEWIS: Why the change?

JEFF: We initiated a redirect and it's moving faster than we expected. We can't change it now that it's on course.

LEWIS: For where?

JEFF: Just tell them to go straight to D-G, fast as they can. Order them.

We didn't question it. We ran a diagnostic on Baby and she was perfect.

"I don't like their excuse," I said. "They've never done a redirect toward a target, only away from Earth."

"It's not an excuse. It's a reason."

"For what.?"

"Why it might not work."

I gave that some thought and said, "We don't have to do what they say, do we?"

Rob stared at me. "What do you mean?"

"What can they do to us? Nothing. We should do what we think will work. The D-G trenches are known to be no good, right?"

"There's an asteroid coming at us, Kit, and they can't alter its course, they just said so. They want us as far from impact as possible so they're sending us to D-G. I'm guessing its for our own safety and has nothing to do with water or reviewed telemetry or anything else."

"And they'll never tell us? Is Lewis hearing this?

"No. I signed us off. Let's go."

As instructed, I first pressed Baby's self-check switch. She not only looked alert, she raised her hand, tilted her head and said, "Good morning. I'm feeling much better."

Crafty Lewis.

Baby was locked down and we took off.

40

On our way, we stayed in contact with Lewis who said to Rob, "You look worried."

"I'm focused. Not much time left and a lot to do. We'll check in later." He signed off.

"You're not telling Lewis?" I said. "That D-G is a diversion?"

"He heard everything they said. He's figured it out. I don't want him to know we have." He looked at me, "Do you?"

"No."

"Is this as fast as we can go?"

I turned all six tracks and we moved faster crab-like across the snow-frost. Our cabin's bubble was alight with head-up displays: our location coordinates, time, both Earth and Mars, and a constantly revising scroll of data as we sped toward Dodo-Goldilocks.

"Do you think they know where it'll hit?" I said.

"No. Do you?"

"No. Let's ask Baby."

I toggled a switch and asked her, "Where will the asteroid hit, Baby?"

Riding out on our deck her lights came on. Her answer was quick.

"Very close."

We laughed. Did she mean dangerously close? Or was it something Lewis programmed in to answer any question about position?

"She doesn't know NASA holds all the cards," Rob said.

"She doesn't need to."

Our new best friend, Baby, suddenly meant much more than safe handling of a drill. She'd just become our lead. She had the equipment, plus the analysis.

"Where is H2O," Rob said to her.

Baby turned her head toward our cabin. It took us a second to realize. Her cameras were focused on our water packets.

We laughed and Rob said, "No. Water in the ground. Deep water. Stop, Kit."

"Here?"

"Yeah."

I did and Rob toggled the lift, set Baby down on the frost.

"Find H2O in the ground," he said to her.

She indicated the ground we were on.

"No. That's snow. Liquid water, Baby. H2O, deep in the ground. She's designed to sense liquid water but I don't know how deep. I want to find out."

Lewis's face was suddenly on our screen. "Why are you not moving?"

"Just trying something," said Rob.

"Get going. Move, now." His voice was urgent. "Straight to D-G, no place else. Don't play around with the bot."

As Rob craned Baby back onto our deck he whispered, "We're right. Asteroid's coming at us."

"They'd never let that happen."

"They just did."

***

With speed wheels lowered we moved so fast that out on the deck Baby began to rock dangerously.

"Why's this happening?" I said.

"Something different since we repaired her." He was popping open the dome. "Keep going."

We were still wearing our skins, remember, and he climbed out on the deck and motioned for me to ease the grips on Baby's cable holds. He tipped her forward till she was lying down, her bronze helmet-face touching the deck. Her body was rattling with every rock or bump we went over. Rob climbed on top of her, looped the cable through his arms and nodded for me to tighten them again. When I did his body was holding Baby down.

"You going to ride like that?"

He didn't answer.

We didn't communicate with Lewis for a while, didn't need to. Cameras all over the track showed him exactly what Rob was doing. Lewis didn't interrupt and we went full speed toward D-G. When I'd ask how he was doing, Rob kept saying fine, great, but I knew he wasn't. He couldn't be. It's a tough ride inside with seats. Outside, spread eagle on top of metal Baby...

41

An hour later we were at Dodo-Goldilocks, Rob inside rubbing his arms and legs, Baby outside cruising around. Her bronze face had a small dent from the rough face-down ride. It gave her a permanent crooked smile. Rob began exercising, swinging his arms, trying to get his circulation going again.

With me in the sleeves, Baby went to several places drilling for a water signal on her own. The readout kept showing liquid ice at ten centimeters. Frost.

"No good for us," Rob said. "But maybe deeper..."

He went out with Baby unable to sit still in the dome with me monitoring everything that came in from Lewis or NASA's delayed comm as well as the data from Baby's laser drill.

We knew for certain by then, dispatching us to Goldilocks was a diversion from the rogue asteroid's path. Lewis was even talking to us about it.

"It might miss completely," Lewis said. "Or it might impact the ice cap, or nearby. That's a broad area, could have included you."

"But it's where the ocean and the springs are," I said. "Maybe the impact will work."

But it would require a direct hit and what they'd done--the out of control asteroid--was slapdash, not a popular scientific method.

After a while, Rob came back in the cabin and we watched Baby moving about slowly taking in data, analyzing, moving on, the perfect Martian worker, unaware of danger, unworried, focused. We were quiet. She kept doing shallow drill tests, the data always the same, negative.

Lewis spoke up from the colony. "It's quiet here."

"Here too," I said.

"We're probably safe," Rob said, staring off at Baby in the distance.

Even control at NASA had gone silent, was no longer asking for input probably knowing we'd figured out that they led us to safety, not water. Or maybe they were transfixed watching the possibility that they might very soon lose us as well as their funding.

My mind was drifting.

"We never used the core drill," I said.

Rob said, "Look, over there, the barchans."

On the dunes in the distance were rows of dark, giant, spidery looking creatures that were actually carbon dioxide geysers like the one that toppled Baby, only much bigger. Rob started frowning as he watched Baby move toward them.

"What's she doing?" He leaned forward. "Where's she going?"

I sat up straight and looked.

Rob was right. Something was disturbing about her going so far from us. In her busy search, she was now headed for the barchan field, a forest of geysers in full blow.

"I thought she could learn," I said, and tried to control her with sleeves but her search-for-water telemetry was an override. She was going right into the blasting CO_2 that would send her high.

Rob leaped out of the cabin and ran. I watched, didn't know what to do. At the same time, a bright, white light came from behind us. Like a searchlight in the sky, it was flooding everything in every direction. The streaking asteroid was traveling over us at 18 km/sec, (that's over 40,000 mph) crossing the sky at such a shallow angle it looked like it might miss the planet completely and slingshot out into space. I stood up and watched. Rob stopped running and watched.

I came to my senses and smacked the control board switches, brought the sleeping track to life, and shoved her into high speed toward Rob. He leaped on the deck and dropped himself inside with me. As he did, the asteroid did not miss Olympus but impacted its caldera bowl top with a shock so strong that even in our thin atmosphere and us holding on inside the cabin, the 2400 kilo track lifted, Baby's grappling hook swinging, and tipped over on its side.

Rob and I lay cradled in the dome, our ears ringing from the impact wave.

When my eyes focused, I was looking through the top of the dome which was now level with the ground. In a few seconds, before either of us had said a word, a rock cracked down on us. Then a few more. Then a thundering rain of rocks that worsened to a downpour of debris, rocks of all sizes and sand and huge chunks of permafrost, so loud it was like a war movie. I couldn't think clearly. It went on and on until gradually it began to lessen and there were only scattered hits and finally, those ended.

Our minds cleared. We rolled our heads, looking around. Sandy rubble covered most of the dome where we were cradled in the sideways lying track.

We couldn't see Baby.

***

We lay there breathing hard, only our eyes moving. The dome had held solid, but was so covered with rocks and sand that only thin streaks of light were coming through.

The voice of Lewis said, "Talk to me. Talk to me!"

We sat up on the side of the cabin and saw his worried face

peering at us.

"What happened?"

"Track tipped over," I said.

"You both okay?"

"Yeah..."

"Baby," Rob said, ignoring Jeff.

We crawled out the cabin's side exit that now faced upwards. Sand and debris slid away, some sifted inside the cabin. We slid down the dome and looked for Baby at the edge of the still spouting geysers.

We made our way through the freshly rained down junk, got closer, and saw her, in pieces.

"Cripes." Rob lifted a Baby arm, completely separated.

Lewis, as he saw what we saw, said, "Jesus."

"Look at this," I said.

My camera's view showed Lewis Baby's computer-loaded chassis lying in the debris, deeply dented which meant the insides were deeply damaged.

Lewis said, "Talk to me. Tell me what happened."

I looked over at Baby's third part, knowing Lewis would see. Her head was sitting on the ground, neatly severed and looking straight ahead, perplexed.

42

Back inside the track, it was agreed to discuss details of the incident later.

"It missed you," Lewis said.

"Yeah, we're lucky," Rob said. "How do we get the track upright?"

"Baby," Lewis said.

"She's in three pieces, she can't upright anything. Things keep turning over, Lewis. You gotta work on that."

"I know...I didn't see that coming. Bring Baby back to the track."

"It's hopeless. We can't rebuild her again."

"Go get her, Rob."

Following Lewis's instructions we removed the wheels and slid the flat, box shape chassis across the sand with one Baby arm, the wheels, and her helmet-head riding on top. That made it heavier and a struggle. Her dented face with a new mark by the camera eye-slit appeared to be giving us a wink, funny if we hadn't been so tired and she weren't so torn apart.

When we got Baby to the track, we crawled in the dome and Lewis told us to rest. We also needed to eat and check in with the infirmary which we did, lying sideways in the dome.

***

"They were riveted," said Chang.

"I can't believe you let them watch."

169

"It worked, Kit. They're quiet now. They think it would have happened to them; Baby hurt again and them probably dead."

"They would be," Rob said.

"They're glad you're safe. It's a vicarious adventure for them. Nobody's watching memories pieces anymore."

"No Fourth of July parades?" I said.

"None. Remember, they haven't heard a word you or Lewis or JPL have said. They just watched the asteroid and you not getting hurt. Very lucky. Very exciting."

"That's good to know," Rob said. "How's my mother?"

"The same. Quiet. She looks peaceful."

Mr. Elkins' head appeared on the screen. "Could I say something?"

"Still guarding the door, Mr. Elkins?" I said.

"No need. Everybody's glued to their screens." He nodded and gave us a thumbs up. "Keep doing a good job. Rob, I talk to your mother now that I'm not at the door."

"You do? Thanks...thanks."

He gave us a little two finger salute and went away.

"He sounds very...cool," I said to Chang.

"He is. He won't tell."

Then Rob said something to Dr. Chang that surprised me and maybe even Rob himself considering everything that was happening.

"Sit with my mother when you have the time, will you? Tell

her I'm off somewhere, anywhere, but that I'm safe and not to worry. I know she won't hear but..."

"She might. I'll do it," Chang said. "Right away."

"Thanks."

We ended the call.

43

With Lewis guiding him, Rob reattached Baby's arm, "Simple." Then they went to work on the chassis and that was "far from simple." Much more serious damage there.

Rob opened it and Lewis took a look.

"Well, no more drilling for Baby. Those relays and chips are..." He trailed off. "Everything else is intact though."

"We can continue the water search?" I said.

"If we can reattach her head."

"Is that possible?" Rob said. "Tell me the truth." No answer. "C'mon, Lewis. What do you think?"

"I think it can be done...maybe. We'll have to lose the last camera eye to do it."

"Blind Baby. That's okay."

"No, it isn't. She'll safe. She'll freeze."

"Override it."

"Can't be done."

"You built the robot and can't override the safe mode you built?"

"Yes. I didn't know what you'd encounter out there. It seemed sensible."

"Okay, then build an override."

Lewis looked up slowly, thinking, and said, "There's a separate...maybe I can..."

He stopped, still thinking.

"We don't need to know how. Just do it."

"What still works in the cabin?"

"Very little and we need it to be upright to get food and water and heat. I don't want to spend a sub-zero night, Lewis. We wouldn't make it."

"Yeah you would in skins. Just...give me some time."

Rob switched off audio and looked at me, nothing we could do at the moment. We climbed out of the dome and lay back on its curved glass just as the sun was setting. When the sun dipped and it was black dark, the stars were brilliant. It was comforting looking up at them.

"What do we do?" I said.

"We're not really stuck," Rob said. "Lewis is too smart for that. He built all of this. He'll make it work again."

Rob was looking at me in a particular way he has, looking at my face but his mind off somewhere else in deep thought.

"What are you thinking?"

He blinked out of his thought, "Come on."

We slid down the dome and walked around the track checking

the exposed underside. Rob crossed his arms on his chest and lowered his head.

"What?"

He looked up suddenly, ran around to the cabin and climbed in, me right behind him.

"Put on the robot sleeves," he said.

Rob contacted Lewis again. "What if we fire her up and switch on robot arms and she doesn't feel like she's blind because somebody else's eyes are looking around for her?"

"Fool override?"

"Yeah, the way the sleeves fool her. She trusts them, knows where to go and what to do without...thinking, wiring. Right?"

"Maybe."

"Worth a try, isn't it?"

Baby's parts were laid out where Lewis could see. The track's lights beamed on the work area. Lewis gave orders, and like a surgical nurse, I handed Rob tools as Rob followed Lewis's instructions. Baby's head was reattached then they went to work on her chassis, piece by piece; circuits, cables, everything, repairing as much as possible.

Baby soon stood before us put together enough to function, we hoped, though her winking eye slit wasn't reassuring.

Lewis said, "Put on the sleeves, Kit."

Braced sideways in the cabin, sleeves on, I activated the system which should have activated Baby, but I couldn't see her from my angle inside the cabin. I could see Rob's face though and it lit up. I raised my sleeved arm and Baby's arm rose.

Rob and Lewis whooped. Baby was active.

"Put her on her own, Kit," Lewis said. "Let's see what she can do, if anything."

I took my arms out of the sleeves and Baby stayed on, but she didn't move.

Lewis said, "Half alive and can't see a damn thing."

To which Baby tipped her head and said, "Good morning, how are you?"

44

Fifteen minutes later as we were working, we heard cheers and applause from JPL and Jeff talking through Lewis's feed.

> JPL: Unbelievable, Mars. You guys are amazing.
> MARS: Yes, we are. Want to know how Baby uprights a 2400 kilo machine? Watch this space.

"They'll see in fifteen minutes," I said.

"Or maybe they won't," Rob muttered, but he was smiling.

We were pretty cocky by then. Tipping over had become something we knew about, though the track is a lot heavier than Baby.

"They're plenty happy," Lewis said about JPL.

"Not bad here, either," said Rob.

It was a party atmosphere, the track lights flooding the area and Baby busy gliding around. It looked like the circuses we'd seen in memory pieces. Phobos made an appearance quietly trekking across the sky as we labored. That must be what a pet is like, I thought, appearing from time to time to watch what people are doing.

Using the sleeves, because Baby really was blind without her camera eyes, I maneuvered her to the track's cabin. She couldn't drill anymore or look for water again, but we only needed her muscle.

"Just lift it up and we'll take you home," I heard Rob say to her.

We were exhausted and exhilarated, for being alive I guess, adrenaline pumping.

"Okay, Kit," Lewis said. "Show her what to do. Lift it and tip it like you did before."

Rob made a shallow channel under the side of the track that lay on the ground. It was less than a meter from my face as I lay sideways at the controls. My human hands in sleeves put Baby's metal hands in the channel and lifted.

Nothing happened. She should have been doing what I was doing, the exact same thing. Of course it was heavier, a lot heavier.

"She can do it," Lewis said.

"Can she?" I said. "She can lift this weight?"

"Should be able to. Keep lifting."

Nothing happened.

"Is she thinking it over?" I said, only half joking since she kept chirping her greeting.

My only thought was, please don't safe.

My arms in sleeves were lifting and nothing was happening. Nobody said a thing. I kept lifting, nothing. More lifting. Nothing. Still lifting and I felt a slight movement, very slight. Baby was doing it. I felt the track rise a few centimeters off the Martian regolith. I continued to rise, slowly, smooth and gentle. I started seeing bright stars out of the cabin's dome instead of barchans. I reached the tipping point. Baby's arms were at their highest reach and she held me there, balanced. Instruments showed the track at 45 degrees. I felt a slight wobble. As I held my breath, the track dropped on over to a solid but gentle, upright landing with me well locked in my seat and dust rising. The track was once again flat and drivable.

Still enveloped in the dust I said to Lewis, "You'll have to build in a self-uprighting apparatus, Lewis..."

"Yeah, I know...concentrate."

Rob appeared through the dust and jumped into the dome with me. "Wait till JPL sees this," he said. "And NASA."

We lowered our audio volume.

***

NASA was watching, of course, and would be 1) glad we were safe and 2) overjoyed to have access to samples from Mars' deep core, thanks to their errant asteroid.

That was luck, unthinkable before the impact. A silver lining, I believe it's called, referring to clouds back-lit with sun as seen from Earth. We don't have silver lined clouds here, hardly any clouds at all, but we admire them in Earth films and photos.

The message from NASA/JPL that arrived fifteen minutes later confirmed their noisy joy about our safety and the prospect of fresh Mars samples.

> JPL: There will be tests we never thought we could run. Couldn't get those with rovers.

Of course they couldn't. We live here and couldn't reach them until the asteroid. We could see mission control was crowded, more senior people there, Carla for one. She was lead on several big projects and was very happy for us.

> JPL: Congratulations, you two. You're heroic figures.
> MARS: You can forget all your rovers. We have an endless supply of material from the guts of Mars. Let the testing commence.

177

Lewis was grinning. That was rare. He was usually solemn in his science world but this trick--building Baby and remotely repairing her in a field of deep, testable, Mars debris (twice!) and then uprighting the track--made him happy. JPL must have perceived that.

> JPL: Good work, all of you. We'll be working on tests together.
> MARS: Hey! Flowing water! Or we won't be working on anything.
> JPL: Carla hasn't forgotten that, but at the moment what concerns her most is your proximity to the asteroid impact on Olympus.

I asked Lewis what she meant.

"I reported ground reverberations," he said. "And she thinks you should get out of there.

"That again?" Rob said.

"I doubt if a pyroclastic flow could reach you before it freezes but we can't be sure how it would behave. On Earth, it's 1000°C and moves 700 kph.

Without a word, I swiveled the track wheels toward Baby. The crane hoisted her on our deck and hoping the locks were not damaged (we hadn't tested them) I flipped the switch. Retaining clamps secured her. We were good to go.

"Move, now," Rob said. His tone was so urgent it startled me.

"You drive," I said. "I've had enough."

He lowered the speed wheels and accelerated even though we were in rocky debris.

Lewis appeared on the screen still looking happy. "I told JPL frozen water's a little peak on the graph right around zero degrees Celsius."

"Think they'll appreciate your sarcasm?" I said.

"Dunno."

Rob wasn't interested in our science humor and asked Lewis what's in a pyroclastic flow.

"Hot gases and hot rocks. Lucky for you they're fluidized. They'll cool and solidify but I don't know how fast. Don't worry, it won't fall on you."

"The track doesn't go 700 kph."

"You don't need it to. Yet."

45

We sped along not talking, homeward bound, Baby stable on the deck. We were exhausted but I don't remember feeling it at the time. It was going to take all night for us to be out of danger.

"Music?" I said.

We were both thinking about everything that happened: Why? Who's at fault? Nobody? Maybe all of us? So I got some music going that seemed right for the occasion. We've been exposed to Earth music. In class, we listened to operas and classical music and rock and modern, and we learned about the composers. That night I chose something with a beat. It got us moving remembering the movie it came from. It's old, but we'd watched with parents, the guy in white doing his dance, his strange shoes, the strange complicated places crowded with people and problems. All Earth things seem complicated to us, guess I've mentioned that. You've survived so much. I was thinking it again for the hundredth time as our speed wheels ground across testable debris, our precious, rained down booty.

Suddenly Lewis was on our screen.

"JPL says what I picked up from Olympus does look like a Richter read-out."

Rob said, "There's a quake at Olympus?"

"They're calling it some kind of movement and flow," Lewis said. "Not a quake."

"What does movement and flow mean?"

"It doesn't read like an eruption. There's no explosion, you'd have felt that. It's rumbling and shuddering, slow-leaking, hot, near-surface material."

Lewis ran JPL's images for us so we could see the MRO view of steaming flow making its way across Olympus's flat caldera.

"You'll be okay," Jeff said. "It's gotta cross 20 km of caldera, climb the scarp, then come down 25 km to level ground. That'll slow it and eat up time. And cool it."

Lewis split the screen again changing to another closer view where our position was marked.

"How far back is it?" Rob said.

"Twenty-four hundred kilometers."

"Jeez...speed?"

"About 100."

"Not 400?"

"Four hundred is down the side of an erupting mountain in Earth-g. Ejecta from the asteroid impact is slowing it even more and making it thick, so think cookie dough. It's not going to catch you."

"Will it freeze?"

"It'll cool and that'll slow it, yeah."

A few seconds passed. "I can't make the track move any faster," Rob said.

"We'll monitor everything, for you. Just keep watching."

"Right."

An hour went by. We watched the distance between us and the speed of the flow.

"Is NASA watching this?" Rob said.

"Of course," said Lewis.

"Tell them not to send another asteroid."

"Roger that."

***

Lewis, Rob and I (and NASA/JPL lagging fifteen minutes) watched the approach of the thick molten cookie dough. The aggregate of rocks and ejected junk rolled together looked like the foamy edge of surf on the virtual Riviera beach only it was larger and unpleasant and gaining on us.

"Ideas, anybody?" I said louder than I meant. "Anybody?"

I know Lewis was scrambling, brain racing and in a minute he said, "Yeah."

"Let's hear."

"Dump Baby."

"What?"

"Dump Baby. She's 400 kilos. You'll gain 10 kph and that's all you need. The flow's freezing and you'll outrun it."

Rob said, "Right."

I said, "Isn't the flow slowing enough?"

"Maybe, but it's too close to call. Dump her."

I looked at the cookie dough closing in and he was right.

Lewis must have seen the look on my face because he said, "I'll build another bot, Kit. Dump her!"

Rob already had the hydraulics gripping her, retaining clamps unlocked. Lift, rotate, release. Baby dropped slowly at an angle and gently settled in the fresh rubble with her quizzical, dented expression face-up, exactly right for abandonment.

My throat squeezed. (A warning: Maybe don't anthropomorphize your bots.)

182

I put on a tough-scientist face and focused on the dark space ahead of us.

46

We were safe, well ahead of the flow.

JPL was proudly monitoring us and NASA would continue celebrating until they fell asleep with sweet dreams of testing asteroid splash and the results they would generously share with the world's scientific community.

Lewis was still with us when Chang's face appeared, screen split, all of us in contact.

"Hello, Dr. Chang," I said, and then, almost afraid to ask, "How much did you see?"

"Everything."

"And the others?"

"They watched on their screens, no sound remember, and I watched in my office with Sylva and Mr. Elkins. When I heard what was going to happen with Baby I cut it off completely, told them I lost contact and didn't know what happened, acted like I was trying to get back on and couldn't do it. Dr. Sylva's with them now. We told them it was because of NASA again, security reasons...we can always blame NASA, right?"

"Right," I said. "Did you feel the impact?"

"A mild tremor. I told them it was good, just the kind of thing you wanted to record and you were glad it happened while you were there." He shook his head amazed. "They bought it."

"Good."

Chang looked exhausted and had nothing more to say.

"You should get some sleep, Chris," Rob said. "Are they sleeping?"

"Oh, yeah. All tucked in and looking forward to your return so they can thank you for intervening and saving their lives."

"I guess that's good."

"Oh, it is."

I told him I thought we'd be back by noon and we signed off from everybody. With searchlights blazing we made good time. I slept, then spelled Rob while he slept.

There are parent stories about falling asleep at the wheel of an automobile, how dangerous it is and how often it ended in death. Death by sleep. All those cars speeding in opposite directions inside their little painted lanes. Very risky. Not appealing. How do you do it? How do you Earth people still do it?

I was sleepy again, heavy with exhaustion. I switched to auto drive and full control radar. We weren't going to crash into anything. Radar doesn't sleep and it doesn't make mistakes.

Nite-nite.

***

I dreamed. Fire, this time.

I'd never thought much about fire probably because it doesn't burn in our atmosphere. It was surely the sight of the asteroid impact and the explosion, white and red like the awful fires I'd seen in memory pieces. I know those pieces are important to the parents but I still don't see why kids had to watch them.

Actually, I do.

It's to understand what our parents came from and also it's the only history we have; parents' history is our history, sort of. Just because we didn't experience it doesn't mean we don't come from it. We have history lessons and watch videos about the different wars, or used to. History turns out to be mostly wars which are always explosions and fire. That night after the asteroid, I dreamed about a memory piece we watched in class with Lewis.

"This is about Bloody Mary," he told us and again we heard about the Tudor family in England, very dysfunctional, no getting around it.

"Mary Tudor is considered the most cruel of all the English royals," Lewis told us. I heard him saying it again in the dream. "If you didn't agree with her religious beliefs, you were burned alive."

The whole class said "Ehhh" and "Ewww," and somebody said, "How did Mary know you didn't agree?"

"Because people admitted it. They wouldn't deny their own beliefs."

"Eeewww..."

That puzzled us. How do you deliberately suffer like that? Lewis had no reason to lie so we accepted that it was true but it sounded irrational, completely crazy. Any kid I know would say any words necessary to keep from burning alive. I know I would. We know what a burn feels like. We have heat to cook with and to touch part of a cooker is to jerk back and yelp. A whole body tied to a stake burning. Unthinkable. I had to turn my face away from the woodcut drawing Lewis projected, a family burning at the stake and looking "heavenward", a word Lewis explained to us.

And this reaction wasn't just me. After the Bloody Mary lesson, most of the kids had bad dreams and talked to their parents about it, trying to square that madness with our own bodies and what we feel and think. How did people do it? So parents asked Lewis to stop teaching that part of history or anything gruesome and he said it wiped out an awful lot and they said for him to do his best and that was the end of gruesome history.

Thanks, parents.

We learned other history, Babbage and his Difference Engine, James Watt and the industrial revolution, the discovery of penicillin, the race to the moon, the giant step for mankind that got us here--good history. (We could tolerate space race explosions, pretty sure nobody knowingly signed off on those.)

So my dream that night brought back the willing self-infliction of pain and woke me it was so disturbing. For a few minutes, I lay there trying to figure out why I dreamed it. We'd just escaped death by red-hot Mars core ejecta. I should have dreamed about something relieved and happy, not Bloody Mary. Then I remembered.

Baby. We'd killed her.

She was solidifying under Mars core junk. She was only machinery made to behave like us, but I could see her cocked head saying good morning and tears blurred my vision. She was human to me. To Rob, too. Then reality shuddered through me.

We still have no liquid water flow.

We're going to die.

47

"Welcome home, welcome back!" Parents cheered us.

They didn't know about my dream and they didn't know we needed flowing water. Stayin' Alive had a whole new meaning for us and knowing it was our favorite piece of music, they had the Bee Gee's blaring as we entered the infirmary.

Rob and I broke into a dance, easy music to dance to, and some of the parents joined in. Might as well let them know we're glad to be there.

"May I have this dance?" my dad said.

Rob handed me over and Dad and I boogied to the beat. I'd seen him and Mom do it but not for a long time. For a minute I felt like I was on Earth having fun after a week of work I didn't necessarily like, only in my case it was work I barely survived...TGIF.

That day was beautiful for all of us, kids and parents, dancing like there was no tomorrow which at that point, there probably wouldn't be. How long had it been, a year at least, since our parents were normal, happy and strong, and not bickering? Much longer than that before we noticed it.

I asked Chang, "Aren't they well enough to join the work rotation so we can get the others out of the biosphere?"

"It's risky. They shouldn't be in low oxygen again longer than a few minutes."

"We can't let the kids keep working there."

Rob was with us and said, "She's right. We're celebrating failure."

"They don't know that," Chang said. "And this is good for them. They're celebrating your safe return. You're heroes. Let it go at that."

Chang walked in among the dancing parents saying a few nice words, working his way through the happy couples, studying them.

***

Okay. Time to hook up with Mission Control and fix our life

threatening problem.

Next morning Rob and I hurried to council with Lewis. He cleared the operations room and we contacted JPL in Pasadena and NASA Johnson in Houston. This had been arranged.

In Houston they were grouped in the large conference room, waiting for our contact, a big crowd with Carla. They were taking our failed search for a target, the misdirected asteroid, and our continuing fruitless search for flowing water, very seriously. They cared about us personally, we knew that, but heavy in the room and looming over all of it, was the possible of the loss of funding and the U.S. no longer leading the ongoing space race.

Carla said, "Good morning," and Lewis answered, "Good evening," our days and nights still reversed. (Our orbit makes our seasons twice as long.)

We knew they had to be sleep deprived and we appreciated their efforts to figure out something for us, which they said they had done.

> JPL: We have something new from the ice on Goldilocks, old samples, new analysis. The ice did have salt in it like we thought. It's briny water. The purer water is freezing out first and pushing the salts up and raising the freezing point. We think it happens because of the calcium perchlorate you've got there. It's probably been stirred up from the subsurface by your activities, rover races, everything you do. In water-ice, perchlorates create liquid

water even at minus 73 degrees Celsius.   For your settlers that's   minus 100 degrees Fahrenheit.   Despite your cold surface, you have subsurface water.

MARS: How much?

JPL: A thin layer, several hours a day, spring and summer.

MARS: We know that. We need flowing water day and night, all year for the biosphere basins. It replaces the oxygeneexchange and fixes our problem. We can't put people in there day after day breathing air like they're at 6,000 meters. It put them in the infirmary with no release date. Now it's happening faster. Anybody working there will be affected quickly. It won't take years and we have to eat. We need workers in the bio permanently.

JPL: Let us work on this a little longer. I think we're on to something, Try to be patient.

MARS: Hard to do.

Carla asked if we had any ideas and Lewis said (I think to be polite) maybe we should do the salt thing and see what we get.

It was pretty clear neither one of them wanted to get into the death scenario. Of course she was aware of it but she was keeping it positive, that's her job. She went on talking to Lewis about how to do the experiment with perchlorate. Rob and I left.

I went back to my hab, found material on Mars perchlorate, sank down on a sofa and started reading. I read for the rest of the day and into the night.

48

A salt experiment. Right there in the colony.

As we worked the next morning, Rob and I were wearing skins and glass and were able to use pickaxes and dig as freely as if we were indoors. (Thanks again, Lewis, for the improved, lightweight gear. Just never thought we'd be using it while searching for water.)

The fact is, we were digging carefully, searching for traces of this melted subsurface water Carla talked about. Our middle-latitude, summer season, permafrost, not deep--water, could be here. But moisture is all we found and it was nothing we didn't already know about.

"We'd have to harvest acres of this every day just for human consumption," Rob reported to Lewis. "And our recyclers are already doing that."

"Not enough for the bio?" Lewis said.

"Not near."

"I thought so."

Lewis told JPL and reminded them we needed "multiple liters in constant flow."

All this got me thinking. What I'd read about the salt the evening before gave me an idea. It had to do again with depth, and I told Lewis.

"We have to get way below the permafrost. At our latitude

there might be frozen water down there in larger quantities."

"And we reach it how?"

"Drop a cylinder packed with perchlorate salt..." He looked away shaking his head at this simplistic idea. "Listen to me, Lewis. Release pores in the cylinder will let out the salt that melts the deep ice and we capture the liquid water. Could be enough at the right depth."

"You didn't find any up toward the pole. Why would it be here?"

"Because we didn't try perchlorate up there and it's summer here and warmer and the perchlorate is already doing its thing."

Rob and Lewis looked at each other.

***

JPL: Is it feasible?
MARS: It sounds like it is. We'll have to build cylinders with release pores. Kit made a sketch. Very simple.

Carla agreed there's nothing wrong with "simple" and it was better than nothing, definitely worth a try. She wanted to know if NASA could help in any way, which was surely code for "try anything you want." Polite, but given they're 34 million miles away with an increasing fifteen-minute comm delay, we said "thanks, probably not" and gave them an order.

MARS: Go sleep. We won't be doing it for a while. We have a lot of prep to do.

Carla responded with, "Good luck," and told us to stay in touch with any and all news. We signed off.

Rob and Lewis were staring at me.

"What?" I said.

"How did you come up with this?"

I wondered when they'd ask.

"I did some reading last night. We just have to get down to the other strata with some salt. The little frost layer up here isn't enough but if we could get deeper..."

"And so you drew the cylinder that delivers the salt?" Lewis said.

"Yeah. Simple. I could see it. A perforated cylinder. Another cylinder inside it with matching perforations. A lift release. Perforations line up, perchlorate salt oozes out into the soil and..."

"Oozes?" Rob said.

"Yeah." They were both smiling and didn't say anything. "So..this makes me lead on the cylinder project, right? 'Cause nobody has come up with anything better."

It did. And although we didn't have a yellow brick road (Oz is our favorite of all films) I'm pretty sure we felt like doing some kick-ass gymnastic all the way to the lab. But we didn't. We are serious scientists.

49

We went to work with a big projected schematic of my drawing on the wall screen. Manipulating the schematic, I demonstrated how it would work and the perchlorate would move into the surrounding area.

"Extremely simple," I said. "Like all the best ideas." I was thinking about Einstein.

He said imagination is more important than knowledge, and he didn't speak until he was four. That must have worried his parents. But I have a theory about that. His left and right brains were battling each other, both unusually strong--one imagination, the other number savvy--neither dominant. That's a double load even for an above average brain with 100 billion neurons and 1000 trillion synaptic connections. That battle must have made Albert develop differently, think differently, read late. But I don't have extreme number savvy. I have a strong imagination. So with some help from Lewis and Rob's numbers, this simple cylinder with a release sleeve could be built. Note: this indulgent Einstein paragraph can be deleted.

There was some discussion about the diameter of the cylinder. Too wide and not all the salt would dissolve. Too narrow and the effect wouldn't be great enough to release a liquid flow, only dribbles. We decided on ten centimeters, a medium width, manageable. The length was tricky. Our coring machinery is sizable. It could handle a great depth and liquid water was definitely at a great depth. So did we dare try 800 meters, pretty much the max?

"Let's go for it," I said.

Agreed.

It didn't take engineering long to make the cylinders. Two 800 meter shafts, eight sections each, simple stuff. We chose an arbitrary location just outside the colony. The summer permafrost softens during the day but the nights are below freezing even at the equator and certainly where we are in a middle-latitude. There was no question we'd have to go deep.

The cylinders were made and the track readied for core drilling in a few days.

***

The evening before the cylinder dig, Rob and I went to the infirmary for a visit. My parents hugged us.

"We thought we were going to lose you out there," Mom said for the tenth time.

"Good thing we didn't go," Dad said. "We'd be dead."

So they'd been talking again, all the parents probably, about us doing their exploring, how brave we were and how smart doing what we did to survive.

"Fearless," Dad said one more time.

"Necessary," Rob said, genuinely humble.

I went mute not knowing what to say or how to be.

I hope on Earth you understand that this uncomfortable deception was necessary. Our parents were peaceful, the worst of the low oxygen effects still there but not irritating them to such extreme behavior. They were mostly weak now and didn't look like they'd ever be themselves again.

That afternoon Dr. Chang gave us his diagnosis:

Severe Hypoxemia due to prolonged low brain oxygen levels.

Irreparable heart and brain damage.

Severe irreversible homesickness.

Exaggerated feelings of loss and uselessness.

(I'm tearing up again typing this. It just wasn't necessary.)

Poor Mom. Poor Dad. What a crap thing to happen to them, to all the parents. I have to put it on the record. Delete it if you must, but people should know that our parents did everything right and then this happened. Now all of us were going to die sooner than necessary and none of us deserved it.

That evening I blurted out, "Everything's going to be okay," as though everything wasn't.

Luckily Dr. Sylva heard me.

"Rob, go sit with your mother," she said. "I think she's beginning to come around."

Rob left quickly but I stayed and so did Sylva keeping an eye on me, worried about my inappropriate remark. She redirected Mom and Dad's thoughts by asking casually if my name was short for Katherine.

"Oh, you know, it wasn't," Mom said, remembering, smiling, and I heard the story one more time. "Just before Kit was born I was watching a memory piece about my family and a kitten I had that was run over by a car. It broke my heart. I was thinking about that kitten when I went into labor and when she was born I held her and looked at her and she was my little kitten, all tiny and lovable. I called her kitten for several days."

"So she's Kit," said Sylva.

Dad smiled and nodded.

Mom said, "She'll always be my kitten. We put Katherine on the birth certificate."

Then Mom cheerfully launched into other things about life in Inglewood and Sylva made notes. When Sylva finished and was leaving, I walked with her.

"They seem so well," I said. "I can't imagine what knowing the truth will do."

"Which is why they can't know. Be careful what you say, Kit, please." I nodded. "How's it going out there, whatever it is you're doing?"

"We're drilling for water...too long to explain. But it might work. Wish us luck."

"Oh, I do. Several times a day."

I joined Rob and his mother. She didn't seem better at all. She was still in her perfectly safe, dreamy world, floating along on her own like our little Mars moons.

50

Next morning we were back to work early, more enthusiastic than ever; me, Rob and Lewis with the track doing the work. It straddled the spot and drilled.

JPL checked in with questions.

> JPL: What makes the perchlorate move out of the cylinder?
>
> MARS: We'll heat the cylinder itself and add hot water to the salt. As it filters down, the ice around the cylinder will melt and that releases more salt and more water is added till the cylinder is empty and the ice around it is liquid.
>
> JPL: A school science project. Got it. How much water do you expect to access?
>
> MARS: No way to know till we do the first one, We can drill in more than one location and hopefully find a vein or a spring. There is subsurface ice here. MRO detected it a long time ago.
>
> JPL: But it's deep.
>
> MARS: We know. Only option is another trip to Goldilocks.

Jeff didn't want that any more than we did and left us to our work.

That conversation began to make me wonder if we could go anywhere near deep enough, if we really would need to go back to Goldilocks or some place closer to the ice cap to reach water-ice or springs. That was impossible, way too dangerous.

I went to visit my parents in the infirmary but first told Chang what we were doing. He admired the simplicity.

"Makes sense," he said. "And it's finished, built?"

"It's built." He looked through a port. "You probably can't see it from here. It's behind engineering."

"Ah. Well it sounds good."

"I want to tell my Dad about it, show him his daughter is a champ but...Mom would be proud of me too."

Chang gave it some thought and came up with an idea that would solve the problem. A few minutes later I was regaling my parents.

"Who's in this contest?" Mom said.

"Anybody that wants to enter, any Martian, which is why I entered. It's an open solar system contest, no entry fee." We laughed. "I think sleeve cylinder and perchlorate salt might win. Rob thinks it will. I don't know what other contestants are doing."

I felt relieved although it was telling another half-truth/half-lie, the kind of half character defect my parents detest.

"It's wonderful," they said, impressed.

Dad said, "Simple. Beautiful."

"Not exactly a breakthrough invention," I said.

"No, no, no," Dad said. "Don't put yourself down. It's a breakthrough if you can reach subsurface flowing water. On Mars? Flowing water? Colossal breakthrough, Sport." He gave me a hug.

I enjoyed their praise, came back to the hab and slept well.

51

The next day, ready to insert the cylinder segments, Lewis sent a message to NASA/JPL knowing they had visuals of us each time the satellites passed over. He told them we were ready to start. They answered that it looked good, wished us luck and went quiet, observing.

The track grappled the first section of cylinder, lowered, tightened and locked it. The same with the other sections. Next we poured in the salt, the perchlorate, and filled the cylinder to the top. (There's plenty of perchlorate in our Martian regolith if you Earth teens are ever asked on a test.)

All the parts in place, 800 meters deep, containing salt, it was time to heat the cylinder with a connection to a solar array. That done, we lifted the interior sleeve slightly. The slots aligned and perchlorate was in contact with the deeper surrounding permafrost.

Hot water was required to get things happening--it really was like a school science project. We spooled out a hose that was connected to the lab. Hot water was added and filtered down. As it disappeared from the top layer, we added more until it was topped off. It was a hot cylinder filled with hot H2O and perchlorate salt.

We waited.

Carla at NASA spoke to Lewis.

NASA: What do you see?
MARS: Nothing yet. Give it some time. Let's see if Kit really has MacGyvered this.

I'd heard that reference. I knew what it meant and it was said this time with admiration for me. Knowing they had their eyes on me, I kept my eyes on the cylinder.

A few minutes went by. Nothing.

Lewis poked at the cylinder's salt and studied it. He added more hot water. Steam rose in the cold. He checked the unit heating the cylinder. He checked the perchlorate again and watched. Something was moving below. He smiled and looked back at us.

"More salt!"

That meant the melting was happening and the brine water we created was escaping into the permafrost or whatever was down there and affecting it. It was working just the way I said it would.

Fifteen minutes went by and we heard NASA applauding. It was all I could do not to turn and take a bow but I acted like I didn't hear and kept helping with the salt and water, a humble champ like my mom and dad.

We kept up the process as the salt continued to dissolve and move into the soil. If any quantity of liquid ice was there, it was being affected. We were changing something. We kept at it until the process stopped. Salt and water no longer disappeared. The brine had saturated as far out as it would go and the cylinder wouldn't accept any more of anything. We knew that would happen but hoped it would go on longer and reach out farther.

Lewis lowered a spectroscope and took a sample from outside the perforations.

Did we get anything? Was all this for nothing? I suddenly wanted to disappear, just not be there if my school science experiment didn't work at all. So glad I didn't take a bow.

We waited.

Inside the lab, Lewis heated the soil, captured the evaporated steam and condensed it. He marched out holding a half-full beaker.

"There's maybe 250 ml here," he said to NASA's watching satellite. "That's it. If we did this day and night there wouldn't be enough to pump through the basins. It's only frost water, deep permafrost."

Control suggested we try other places.

> MARS: Okay. We'll move our set-up and do it again. It'll
> keep us busy while you work on it.
> JPL: You could hit something.
> MARS: Not likely in permafrost, you know that. We're
> living on it. We're familiar with it. We were hoping it
> would be shallower than we thought or in a different
> configuration under us, that's all.
> JPL: Maybe it is. The MRO can't see things like that.
> Keep trying.

NASA sounded concerned but I think it was desperation we were hearing from Jeff. And empathy. I walked away and Rob caught up and put his arm around me.

"It worked," he said. "You didn't guarantee what we'd find."

He was right. And NASA and I and everybody else knew we weren't likely to hit flowing water here. It's in the form of ice and much deeper, an ocean and springs under the pole. The *pole*.

Rob and I went to see my parents and something awful popped out of me.

52

"Including kids," I said, "we're down to half our workforce."

Rob's eyes went wide.

My dad was arriving back from the infirmary kitchen with lemonade for all of us and I repeated it for him, dangerously near telling the whole story: We betrayed you, we need flowing water or we'll starve.

Rob interrupted me, "We don't need to talk about that, do we?" his eyes still wide, warning me.

We sipped lemonade and Mom watched us with a slightly bewildered look, childlike, a look she had much of the time by then, lost but happy knowing the grownups were taking care of her. And Dad was becoming more and more like her. He seemed afflicted too, both of them regressing like the other parents, to pleasant children. We went on talking for a while, no more dangerous words from me.

Later Rob and I found Chang and talked to him privately.

"Look," I said. "We've got to tell them."

"They wouldn't understand. They think it's some minor issue in the biosphere that's keeping them from working. They all do. They never question their weakness, never mention going back to work. Their brains are damaged and..." He stopped.

"And what?"

"It appears to be permanent."

Rob's eyes went wide again. "You mean they're as recovered as they can be? This is it? They won't come out of it?"

Chang nodded. "They still do the trampoline almost every day unless there's wind and heavy dust, but they're not getting better, none of them."

"They're like children," I said.

"Exactly."

Rob said, "They're not unhappy?"

"Not anymore. They're peaceful and they're proud of you two for doing the exploring and saving them from sure death. That's what occupies them. And they're excited about the new Baby Bot. They make us take them to see her every couple of days. They talk about her like she's their child."

I suddenly understood why I blurted out that we're losing our workforce.

"The secrecy's killing me," I said. "I have to tell them."

"Why?"

"It bothers me more than running out of food. It feels terrible. We've never lied before."

"We never had to," said Rob, hugging me, knowing what I was feeling.

Chang said, "Right. Listen to me, Kit. The fact is you've had an easy life, all you Gen2s, and now it's going to be like Earth: who gets to work outside the biosphere, who gets to stay alive, a competition. Everybody will have to take turns and everybody will be trying to avoid it, doing whatever they have to do to get out of bio duty."

"Competing for survival like Earth?" I said.

Chang nodded.

We hadn't forgotten Chris Chang was a settler and like Lewis he understood both Earth and Mars worlds and how easy this one is. That made what he then said undeniable.

"Everything will change."

***

It may sound childish, but try to understand. At that point Rob and I needed a distraction, it was a long time since we'd had fun. Our parents had the trampoline, we had sailing through the air on dust boards, so that's what we did. The minute we climbed that first dune and took off we were relieved. The problem wasn't cured, of course, and we didn't have any new ideas, but we weren't trying to think of any.

Instead, we soared and drifted and looped down slopes then went home to my empty hab and spent the evening watching skateboarders on old videos, one more time amazed at Earth kids. Those teens in heavy Earth-g sliding the edge of their boards along the edge of any available rail or bench or low concrete wall, flipping the board under them when they sailed off then landing on it again...impossible. Apparently not. How do they do that? (You can delete this skateboard part but I hope you don't.)

One of the skaters was Canadian and was winning a competition.

"Guess they skate in Canada," said Rob. "I thought it was

cold up there all the time, not outdoorsy..."

I nodded and after a minute realized Rob had gone quiet. I looked over at him. He was staring off at a corner of the hab.

"What's wrong?" I said.

He was deep in thought so I waited.

He turned his head, looked at me and said softly. "...the High Arctic..."

"What about it?"

"Have we forgotten? Remember studying Axle Heiberg Island in the Canadian High Arctic?"

"Yeah."

"Way below zero in a region of year-round permafrost."

"Uh-huh."

"There are three springs flowing there all year, constant temperature and flow rate."

"Right, but that's Earth..."

"...same metrics here. Under permafrost, sub-zero temps..."

"...but not exactly the same."

"Close enough."

"But, Rob. Tapping springs is what we've been trying to do and..."

"...and I just remembered Axle Heiberg and I know now they're there. I really know it. Even Earth scientists thought it meant springs were there, exactly where we were, up there where the mounds are. We saw them, like the ones hot springs leave on

Earth. Doesn't that make a difference knowing that?"

"Yeah, kind of, but how do we get up there again? Olympus is still seeping and that cookie dough is still moving."

"That doesn't have to stop us."

"The flow is a thousand degrees."

"Only when it blows out, not by the time it gets to where we'll be. We watched it slow down and solidify. And we'll be hundreds of kilometers away. It's not flowing fast at that point."

Rob moved close and squarely in front of my face he said, "Look. The fact that we've done it is exactly the reason we should do it again. We know how. We know what to watch out for. We're safe. We know we can get back here if it erupts. We know how to dodge around a flow which is only seepage now, not like right after the impact. Those springs are there, Kit, just like at Axle Heiberg. They have to be."

I didn't say anything. I didn't have a rational objection.

"Don't be afraid. We've done it once so we'll be better at it. Let's go search for springs we know are there." I hesitated. "A high arctic ecoclimate, same as on Earth, exact same geophysical position. On Earth they're flowing. They gotta be flowing here too." I didn't answer. "Kit...?"

"It's crazy."

"No crazier than your cylinders."

He had me at cylinders, but he didn't know it and said, "We'll all die for sure if we don't go..."

"...so it's the only sensible option?

"Yeah."

We laughed, really laughed and it felt good.

"If we die, we'll die together," I said, or maybe he said it, I don't know. We were both laughing.

53

We started preparing that day.

We didn't plan on telling NASA/JPL until we were up in the cookie dough again and maybe not even then. They didn't follow every move we made out of the colony. And if they did they couldn't do anything but watch, delayed.

"We'll leave under cover of night," Rob said. "It's the only way."

"Right, but what about the parents when we don't visit for a couple of days?"

"We should leave them a note so Chang and Sylva don't have to invent one. Both doctors should know, so should Lewis."

"Right, but only them."

We agreed.

We composed a note to Chang and the parents using Rob's all or nothing reasoning, a great adventure, bearing in mind the note might be the last thing they heard from us.

The note to Lewis was longer. It went on a bit about tapping into a polar spring but didn't deal with how to keep any surface flow from freezing before it could get to the colony. Neither of us was dealing with that. There was a good chance it would be hot springs so maybe it would flow steaming all the way. If not...

"Put salt in it?" Rob said. "To keep it liquid?"

"Sure."

"Then distill it in the colony?"

"It's gotta be a hot spring boiling from so deep. That'll be a small detail once we've tapped into a gusher."

Saying gusher made him smile.

We hid the notes inside our skin gear hanging as always at the airlock door.

This felt like a TV mystery our parents might have watched, dangerous moves in the night. Those mysteries were always confounding to us with their complications and our lack of terrestrial context. But our dangerous moves were not complicated. They were logical and the context was simple: No flowing water? No more life on Mars.

Planning this on our own, the two of us were in control. If we died we'd be together. Worst case scenario, we'd fling back our face glass and breathe; chilly but quick.

I felt confident.

Then we realized we couldn't jump in the track and tear off into the night. The track needed to be in perfect condition, stocked with food and supplies. And we needed a new Baby Bot. We couldn't drill without her. She'd have to take the risks like before. That's what robots are for.

***

Lewis shook his head. "Crazy idea."

"We were going to sneak off, we even wrote you a note but we need your help."

"You need to think straight," he said.

"Okay, here's some straight thinking," Rob said. "If we stay, we all die. If we go, there's a chance we won't."

"It's suicide."

"No, it isn't." Rob gave him the Axle Heiberg example. "And since there's every indication the same kind of hot springs are at our pole, it's suicide not to go."

"Those springs have to be very deep."

"Maybe not now that Olympus is seeping. He's disturbed. He's got a belly ache. Things are churning and moving, maybe even the springs. Who knows? It's easily worth a try. You do realize it's suicide if we don't go."

Lewis stared at us.

***

Baby building went faster than the first time. She soon began to take shape while the track got a full overhaul plus the addition of extenders that would prop it upright again if needed. The same was added to Baby, taking no chances. We began to realize that first trip was a test run, nothing more.

"Anything you want special this time, a clock that runs, maybe?" said Lewis. We looked bewildered. "Old Earth joke."

Lewis joking? So he was into this too.

Yes, it was a life or death adventure, but we were way past that. We were ready to go, could hardly wait to leave, then Lewis made us slow down. I still don't know if he did it hoping we'd change our minds but at the time he said it was for the good of the colony, for the morale of the parents.

He refused to finalize Baby.

"There has to be another Baby Bot," he said. "They know I'm building one. When she's gone what do I tell them? You took her to maybe save their lives? That again?"

"Yeah," we said.

"You can't do that. They need a bot that's theirs, for here, for their company as a pal. They're almost children again and Baby's their friend, personal to them."

The track stood ready, supplied with everything we'd need, in perfect working order. We wanted to leave that night.

"Can't do it," Lewis said. "Can't release your bot until another bot's completed. It'll only be a shell, it won't take long."

There's no arguing with Lewis.

So for 18 hours, the colony's food supply continued to

diminish while the engineers worked on Baby Bot 3. Her arms would move with sleeves and she would occasionally greet them, "Hello friends, how are you?" but she had no high-tech innards at all, no water sniffing or drilling skills.

It was two full days later when everything was ready and our departure was set.

Lewis would see us off in a pitch black Mars night.

54

Tiny Phobos gave off little measurable light though he was scooting along up there. No lights were turned on outside. The only light was inside the track. No one saw us. No one could hear us. (How is secrecy possible in Earth's sound-bearing atmosphere?)

We rolled away into a beautiful night for travel, not much wind, very little dust, brilliant stars and Phobos. Despite the undercarriage of the track being taller than most rocks, we again used speed wheels and made good time not yet going across cookie dough. That was several hours away and still farther to snow-frost. Neither of us wanted to sleep. Too excited. Scared too maybe.

With mostly smooth, red dust and small rocks around us we made ourselves comfortable and pulled up something on our screen to watch for entertainment; an aerial view of New York

City, not live of course. It was from the years when my parents were in Inglewood and "often drove across the bridge to Manhattan."

We'd seen it before but it still dazzled us, the streets thick with cars and buses, the sidewalks jammed with people. The time on a building showed noon, a busy time with people going to eat somewhere in the middle of a work day. This had been explained by nonchalant parents as we watched, amazed at their skills in negotiating all that was happening around them.

"How do they do it?" Rob said in the track that night.

I shook my head.

We watched the traffic signals that let them walk or not. And the thousands of kilos of cars right beside them while they continued to talk and laugh like it was nothing, totally normal, which I guess it was to them.

"And the signs," I said. "All those signs in store windows, and screens up high scrolling..."

Why we'd chosen this to watch I don't know, thinking about our parents probably. Once again it was baffling that they or any human could manage that much input every day and survive.

"They must think we're weaklings," Rob said. "Not in a bad way but like we think they're crazy to miss Earth and we don't really think they're demented."

"Right. Think they'll ever understand us not wanting to go there?"

214

"No."

That was a hellish thought to all Gen2s. What an awful, uninhabitable place Earth would be, noisy, dirty, outrageously dangerous. And our parents missed it. They were our never-ending lesson in the comfort of familiarity.

"Oh, well," I said and straightened and looked out at the surrounding area, clearly not dangerous, just smooth dunes and rocks as far as the eye could see.

In fact, it was much more dangerous that night than the New York streets but as far as we could see there was no indication that Olympus would do anything perilous as long as it was seeping. It was wounded by the off-course-because-it-never-had-a-course asteroid, but not fatally wounded. And Olympus must have known a lot of asteroids in its billions of life-years. Maybe not a direct hit like this one but absorbing an asteroid is nothing new to any planet.

Then, as we talked and time went by, we began to see the wide outline of the broad Olympus caldera just visible, creeping above the horizon. We stopped talking and watched, our heading set straight toward it. No chance or desire for sleep now with that in view.

We knew our parents in their childish senility were sleeping well, tucked in, everything calm. I was no longer anxious about them. With us, near the pole, it was a different kind of calm. More like readiness.

***

We began to see the ancient pyroclastic flow lines like brain circuitry across the bald, dust surface ahead of us. We slowed. We hadn't yet reached the cookie dough. We studied what our brilliant floodlights showed us, the recent flow lines solidified and frozen. We got closer and could see they were not thick. Closer still, they looked as though they had stopped moving at all.

We stopped the track and got out. We stood beside the end of a flow line, touched it with a boot. Solid, no steam. It had completely cooled.

"Hasn't been moving for a while," Rob said.

He crouched and tapped it. He took a sample for his wet lab. NASA would go crazy with this, the innards of Olympus Mons that we could scrape off the regolith, completely different from anything we'd studied at our latitude or anything their landers ever scratched up and analyzed.

"Where do we start?" Rob said.

55

We had data from our satellites and knew what they saw and where the subsurface sea was, so deep it was untappable. It was springs we were hunting for and there were spring mounds all around us.

Rob stared at them.

"Like the ones on Earth," he said under his breath. "No wonder they think hydrothermal springs are here."

But we knew that everything on Mars, like on Earth, has moved subsurface, so the actual springs could be at another location by now and deeper for sure.

"Give me a hand with Baby," Rob said.

I pulled on the sleeves and Baby searched the ground for a promising spot to drill. We were at the lowest reach of flow lines. They were at each side of us. We started drilling.

No cylinder and salt affair, this was laser hand drilling, Baby Bot 2 the stabilizing force as the drill went through permafrost. She could handle it. Human arms couldn't going through permafrost this deep and who knows what under it. Inside the cabin, I monitored the read-out from Baby.

"Twelve kilometers," I read off to Rob.

The laser meant we could drill in many places, laser-poking around Mars' deep subsurface for a water signal.

Didn't you use diving rods on Earth for this and you were surprised they worked? We know magnetics as well as you do, so the success of divining rods isn't surprising to us. But could anything, I wondered, even Baby's sophisticated telemetry, divine something as deep as a spring? I was drifting again.

I blinked, remembered I hadn't slept, and felt a little light headed. Not a good condition to be in handling the sensitive controls of a laser apparatus even though Baby was out there running it.

Rob waved for me to stop the drill and climbed in with me.

"I think we should move on."

"Why?"

"We're pretty deep and springs are closer to the surface nearer the pole."

"Let's go."

That's how simple our thinking was at that point: a guess, an idea, made on sleep deprived, low-level uncertainty.

Baby was lifted onto the deck. Rob did a walk-around and checked everything on the track and on Baby. Training: no matter how tired or discouraged--particularly if we're tired or discouraged--do checks and double checks.

Done.

Rob patted Baby's hand. "Good job."

Baby said, "Good morning."

***

Very soon it really was morning and our contact with Dr. Chang was through a groggy, half asleep haze. Lewis was with him.

"We're okay," Rob said. How's everything there?"

Looking haggard, Chang said, "Not great."

"What happened?"

"They're not taking it so well."

"You told them right away?"

"It came up somehow."

"We're on another adventure," Rob said. "They can understand that."

"That's what I told them but they're worried about you, don't know why you didn't come say goodbye if it's just another adventure. They've figured out that it must be dangerous and they're in a state."

"What kind of state?"

"Upset. Angry. Some of them are in tears. I called Lewis to help Dolf in case they got...rowdy. It's not good," he said.

Lewis looked in on the edge of the screen.

"We can't come back now, Lewis," Rob said. "We're just getting started. In fact we're moving on toward the pole."

"You've drilled already?" Lewis said.

"Yep. Drilling went okay but no sign of anything."

"Then why leave? Keep drilling."

"We've gotta get closer to the pole. Better odds."

"More dangerous."

"Maybe not. The flows are solidified and cooled. We're between them. We're okay."

Lewis took a deep breath. "All right."

Chang said. "We'll tell the parents we talked to you and you're sorry they feel bad and you'll bring them a surprise."

"Yeah, we will. Another Baby Bot proving we prepared to leave secretly. That'll cheer 'em up."

We got to the new location fast.

56

"Are you falling asleep?" I jerked up my head, opened my eyes wide.

At the new location, Baby was out on the surface with Rob when his words punctured my unconscious.

"Yes...wow. I think I was," I said.

"We'll have to stop then. You're running things and Baby's arms aren't doing anything."

"Aren't they locked on the drill?"

"Yeah, but the sleeves know if you're falling asleep or disabled somehow. They monitor your heart...didn't Lewis tell you all this?"

"No. I don't think he thought I'd fall asleep."

"It's okay. Eat something."

"No, I'm fine. Let's keep going."

He looked at me hard, serious, standing with Baby--good scientist, safe scientist--then he said, "Okay. Let's drill."

Baby said to him, "Good morning, how are you?"

"Can't you shut her up?"

"It's built in."

"I know...power the drill."

220

I did.

Rob was irritable. I was half asleep.

***

The drilling gave us energy. We could hit a vein at any minute or thought we could. At this location, it wasn't impossible. As the sun rose higher and blazed a pink white reflection around us, work was more difficult. We were drained. Rob noticed it and admitted to it first.

"I'm coming in."

I took my arms out of the sleeves. As he climbed in I pulled out food packs. We peeled back the seals and without a word, ate, looking out the cabin's dome at the expanse around us. We chewed and swallowed. We drank Lewis's concoction of power water, "plenty of amino acid," and I leaned my head back against the edge of the system board. Through my eyelashes I watched Rob do the same thing. I could hardly hold my eyes open but before they closed I saw Rob had gone to sleep still holding his food pack with both hands in his lap.

The next thing I knew, Lewis's voice was coming at us again.

"Hey. Wake up!"

I opened my eyes.

Rob's head bobbed and straighten. "Uh...yeah...we're here, sleeping...first time since..."

"Since the last drill?"

"Yeah."

"Then you've got to get more sleep. Sorry I woke you. Get on the pull downs. I didn't know you..."

"We need to drill. Higher."

"No, no, no..."

"We've got a feeling about this, Lewis. We're going higher."

Rob was already at the controls lifting Baby back on the deck over Lewis's objections, shaking his head. He said he had a feeling about going closer to the pole and the possibilities there.

Dr. Chang appeared on the screen, "You should listen to Lewis."

Rob ignored that, "How's my mom, Chris? Does she understand anything about this?"

"No, I don't think so. She's weaker. More reason for you to be careful and not go farther north."

"For my mother?"

"For all of us. Keep drilling where you are."

But Rob wasn't going for it. He signed off and got us ready to move on. Baby was on deck and locked down, drilling equipment stowed. The track was ready to move which is what Rob reported to Lewis and added over his objections, "We're on our way."

57

At the newer place, closer to Olympus, with steaming seepage all

222

around us, we drilled until noon. Lewis left us alone for a couple of hours. The drilling went the same; nice, deep, dry holes. No water signal from the laser.

"A duster is what my dad called dry holes," Rob muttered.

"Your dad?"

"He invested in oil drilling on Earth a couple of times. Only got dust."

"We don't even get dust," I said.

Rob gave me a withering look. We didn't give each other looks. That was parent stuff and it hurt me; surprised me too. It had to be the lack of sleep I thought, each of us handling it a different way. Then I remembered Rob's mother and how he felt about her.

"Rob, sorry I..."

I wanted to make things right, fix my clumsy remark but my arms were in robot sleeves and he'd gone outside with Baby. He could hear me and we were doing exactly what we were supposed to, working together, drilling deep between not yet solidified cookie dough toward hoped-for water, so I let it go.

Lewis's face on our HUD: "How's it going?"

Rob: "Our chances are better here. Should be. The flows aren't solid yet so they're more recent. We have to be closer to the springs."

"The flows aren't solid?" Lewis said.

"Still steaming, not moving fast."

223

"Rob, listen to me," Lewis said. "You have to come back."

"Why? We're getting close, I can see it, I know it."

"You're too close, way too close. Olympus isn't quiet yet if those flows are recent."

"It's just seepage, leftovers, maybe."

"No such thing. You're being crazy."

"I'm not crazy. There are spring mounds all around."

"Rob."

"Yeah?"

"You've got to..."

Rob tapped a face-glass control that turned off Lewis.

I said, "It's dangerous being out of contact, Rob."

But Rob huddled close to Baby monitoring her moves with the drill. Lewis went on talking to me.

"You've got to convince him. It's dangerous where you are. You're near a volcano that's active, okay?"

"You knew we would be."

"I didn't know you'd be that close. You've got to make him leave. It could blow."

I believed Lewis. I heard the fear in his voice and saw it in his face. That was new. I talked to Rob.

"We really have to leave." He didn't look up. "Rob." Still nothing. "Rob."

He kept working as I sat watching from the cabin. I couldn't see any fresh flow but those would be up higher so I said nothing more. Rob and Baby did the work for another half hour with no results then Chang came on the screen, calm and quiet.

"Hello, Kit."

"Hello, Dr. Chang. How are things there?"

"I need to talk to Rob."

"He's outside with Baby. Drilling."

"I see. I need to talk to him. It's important."

"He won't listen."

"It's...his mother."

When Chang didn't say anything more I knew what it meant. If she was awake and wanted to talk to Rob, Chang would be smiling.

"Kit?" he said.

"I'm here."

"Get him for me."

58

We were sitting together in the cabin when Chang told him. Rob lowered his head slowly until his chin was almost touching his chest. Chang had made us pressurize the cabin and take off our gear. Rob was sitting very still.

"Was she...uncomfortable?" he said, still not looking up.

"No. She went to sleep and never woke up. She was peaceful."

Rob looked up quickly. "Well, we don't know that, do we? That she was peaceful or if she could hear or understand? I think she heard everything but she felt locked out like my dad and was floating out there somewhere with him, watching us with no way to reach us..."

Rob turned away from the screen. I went to him and put my arms around him. He wasn't crying.

"It's okay," he said.

But he had nobody now except me. And the colony. He was calm.

"She's been gone a long time," he whispered.

I heard Chang's voice. I'd forgotten he was watching and I looked up at the screen.

"What did you say?"

"I was talking to Rob."

Rob looked up.

"Come back for your mother," Chang said. Rob shook his head. "Your mother would want it."

"No, she wouldn't."

"I'm sure she would."

"She wouldn't, I know it. You're just saying that to get me to come back. If you really knew my mother you'd know she wouldn't want me to give up and come back. She was a rebel. She was a gentle rebel and an artist, a free thinker." Tears jumped to his eyes. "She'd be proud of me being here and even prouder of

226

me staying. You're using her to make me come back, Chris. Bad move."

Rob stood and said, "Put on the sleeves, Kit. We're going to work," and he went outside.

As I put my arms into the sleeves I looked at Chang and said, "You won't change his mind."

***

We worked right through our exhaustion into the afternoon. We took a break, too tired to talk. As we ate I wanted to tell Rob I was proud of him, like his mother would, but it sounded petty and I didn't do it. Actually, I was proud of both of us. Mostly I was tired. We went back to work and after almost an hour it started.

Sitting in the track it was barely noticeable. Then I noticed it again, a floating feeling in my seat at the controls. I thought it was my exhaustion making me light headed and drifty, then it happened again.

I looked out at Rob and Baby. The drill was still going but Rob had raised his head to me. He felt it too. As we looked at each other, waiting, there was another tremor. I saw his body jiggle. His mouth opened but he didn't say anything. Then a small jolt and a rumbling.

Rob give me the cut-off motion. I shut down the drill and took off the sleeves. Rob was back in with me.

"What is it?" I said.

"Must be an eruption. Let's go."

I flipped switches and swung Baby back on the deck. The juddering made locking her down difficult, but not impossible. Rob strapped himself in his seat and I crab walked the track toward home fast as the cookie dough would allow. The jolts going over big rocks boosted the track's twenty-four hundred kilos off the ground.

In seconds Lewis was on the screen. "You feeling this?"

"Oh, yeah."

"You can travel faster without Baby remember."

"We're way ahead of the flow. In fact, I don't see any flow."

"There isn't any. Not yet. It's a quake."

"What?"

"An earthquake. Have them all the time in California. A marsquake."

"You sound happy."

"Only because you won't die in pyroclastic flow."

"Just in a quake?"

"It won't kill you..."

As he said that, we jolted so hard my head hit the cabin dome.

Lewis's voice said, "Strap in! For God's sake, Kit..."

How did I fail to do that?

Rob went on talking to Lewis, "What will we die of?"

"You won't. A vehicle is the safest place to be in a quake."

"You could fool me."

Looking at the screen showing our entire area I said, "There's no flow."

"How long does a quake last?" Rob said.

"A minute or two. Less."

"It's been longer than that already."

"No, it hasn't. It just feels that way. First marsquake since the asteroid and that was just the ripple from the impact, not a quake. There's gotta be a lot to resettle in the core."

"You sound unruffled."

Another very big jolt lifted us but we were well strapped in.

"Okay, that was a big one," Lewis said, still sounding calm.

"Why are you so mellow?"

"Because you won't die." He was a scientist and we had to believe him.

We held on. It gradually went quiet, no more jolts or shakes, only a slight, continuing vibration that I was surprised to feel in the track.

"That's it?" Rob said.

Lewis said, "For now, yeah, probably. It was..." He consulted a lab instrument. "...just under a minute."

"You're making that up."

"Nope. Fifty-eight seconds. You'll be getting some aftershocks."

"What's that?"

"Shocks after."

"As strong as what we just went through?"

"No, no. Just settling."

Rob was smiling. We'd made it. Then he stopped smiling.

"Still no water flow," he said. And his mother was dead.

***

Back at the infirmary, Chang was handling it.

"Just a quake, like home," he said to the parents. "Enjoy it."
And they did.

"A dust storm is worse than this," one said, and their beds
jiggled with the temblors.

But they were aware that the colony's domed, virtually one-
piece structures are quake-proof.

Mr. Elkins, however, always in the loop, knew exactly what
was happening with Rob and me, that our mission was aborted
and no water was flowing. He sat with his stick, on watch at the
airlock door as usual, grave and thoughtful, while the others
enjoyed the bumpy fifty-eight seconds.

59

The next morning we were safely back in the colony and rested. I
was still dozing.

Rob and Lewis were behind the lab with Rob's deceased

mother in skin gear, the glass locked down over her face. They'd fitted her with a jetpack and taped her arms to the controls. Rob told me his mother's face was peaceful, that she hadn't looked that serene for a long time.

They locked the jetpack into full power and pressed the throttle. His mother's light body lifted off quickly heading straight up, fast gaining escape velocity, much lower here than on Earth.

"It made a tiny contrail and then nothing," Rob told me. "She'll soon be orbiting."

We agreed we'd do that for each other when we died, make sure we were sent back to where we came from, big bang rubble, star dust.

Later at the council building, Jeff was talking to Rob about the danger we were in.

JPL: What were you two doing that close?

MARS: Looking for water flow, Jeff.

JPL: Without us?

MARS: Without you. There are springs up there.

JPL: Right. But you can't go near Olympus again.

MARS: We get that. We won't. We're in damage control now.

JPL: How are they doing?

MARS: Okay for now. Have to ease them into the idea of dying.

JPL: We know about your mother, Rob. Condolences is the only thing I know to say and it's not enough, I know that.

MARS: It is enough and thanks. We sent her out on her own. She's orbiting in skin and glass and not underground like my dad.

JPL: You sent her into orbit?

MARS: We did. Souped up jetpack. I don't care what the other parents want. I know for sure she'd want to float with the stars.

Jeff went quiet--the loss of Rob's mother, the possible loss of us and the Mars Project that had gone on for so long--he just didn't know what to say. Rob understood and tried to ease the situation.

MARS: So. The first marsquake. First since we've been here. What do you want to know?

JPL: Many aftershocks?

MARS: Yeah. Fewer this morning. Lewis has the data.

JPL: Still could be more.

MARS: So we're told. Annoying. (Long silence) You know what, Jeff? We gotta go.

JPL: Sure. Okay. Be careful.

***

We went right back to work. Research.

We jetpacked up toward Olympus--jetpacks travel way faster than a track--and we could see where the flows stopped and even where we turned around and started back.

"We were close to water," Rob said.

"How can you tell?" We were circling our track's turnaround spot.

"Because we know it's there and look how close we were to the pole."

We took a reading on the edge of the flow and sent it to Jeff at JPL.

Rob said, "We could drill a lot of holes and never hit it."

"You giving up?"

"No, just..."

Rob sounded defeated. That surprised me but I didn't say anything.

Back in the colony at my hab, I cooked but we couldn't eat much. No appetite.

We went to the infirmary where the parents were all over us again with praise and joy and with sympathy for Rob. Nothing was said about his mother's non-burial, going into eternal orbit. Maybe they didn't know. After they settled down and we'd talked to my parents, we huddled with Chang and Lewis.

"What do we do about them now?" I said.

Chang said, "We have to tell them but..."

"No buts," Rob said. "You can't just run out of food and then announce it."

"But it would reduce the time they spend worrying."

"That's no alternative. Maybe we should go up and drill, keep drilling, keep trying."

"That's not an alternative either. They'd worry about you the whole time and if they lost you..."

Nobody said anything for a few minutes, Lewis still silent.

Chang said, "Any of us thinking about, you know, assisted dying?" There was no quick response. "Take things into our own hands, help parents and ourselves in the end."

Lewis gave him a strange look that I didn't understand.

"Do you realize we're talking about the end of the colony?" I said. "And all because of badly cured concrete, an avoidable mistake."

Rob looked at me and said it, the awful truth, "Yeah, that's right."

However inexcusable, it sank in.

I said, "We have to let the parents know what's going to happen."

Lewis's strange look was now directed at Chang.

He was about to speak but Chang ignored him and said, "We can give them a choice with assisted dying. We have the drugs and the know-how. They might even prefer to do it on their own, you know? Just...step outside."

Lewis stood and said, "We have to talk, the four of us. Someplace absolutely private. The lab."

60

Only a few lights on, no one else in the half-dark room, Lewis started, all business.

"The entire colony doesn't die," he said.

Rob said, "What do you mean?"

"You two don't know about this, no Gen2s do. There's a NASA protocol for this situation. Total loss of the colony would be devastating for Earth while they're going through their floods and for the space program at any time." Lewis glanced at Chang and went on. "No matter what happens, somebody has to survive--protocol for any space project, always has been."

Chang said quietly, "Sure it's time for this, Lewis?

Lewis nodded. "Word from NASA."

"Ah." Chang lowered his head, looked to the side.

Rob and I were baffled.

Lewis said, "With astronauts in any space mission, at least one comes back, even it means the others have to make it possible."

Rob and I still didn't get it.

I could see Chang knew. We waited but Lewis didn't say anything more.

"What do you mean?" I said.

Then it struck me, how the winnowing of survivors gives the

best chance to keep the others alive.

"You mean...like..." I whispered it *"...the Donner pass?"*

"Not in our case. We can freeze existing food, enough to last months if our number is lowered."

Rob said, "How do we lower our number?"

"Volunteers," Chang said. "My guess is once they know what's going on, most of the parents will gladly step outside. With nobody guarding the door, they'll just...go. You know how many want to."

I rocked back like I'd been hit.

Rob said, "You think they'll just march outside knowing it's easy?"

"Sure. Two couples already fought off Mr. Elkins. Once they remember what they signed about survival, they'll step outside. We'll have food to last till supply probes can get here."

It was quiet then Rob said, "How many would we have to lose?"

Lewis looked at Chang who said, "I'm estimating...about twenty."

Lewis nodded.

I stood suddenly, couldn't catch my breath. "My parents..."

Chang said, "They won't volunteer, Kit. You're Dad's not even..."

"They will, both of them. They're all about doing the greater good. They'll be the first to go, just to get it started...I know them..."

I was crying. Rob held me and rocked me and whispered, "shshsh..."

The meeting was over. Chang and Lewis could do nothing more and left quietly. Rob let me kick the wall of the lab hard as I could as long as I could, and then he held me again.

***

That evening, completely calm, I went to my parents in the infirmary--a test of my bravery. I already knew about theirs. I sat beside them, cheerful, and asked them about things they'd like to recall.

"Tell me about your Inglewood house in detail."

Their eyes lit up. I'd never asked this before, never really been interested.

"Why?" Dad said.

"I'm thinking of sketching it. (Not a lie.) I'll add sketches to the memory book you gave me. Fill it out, make it more complete."

That simple request gave them such pleasure I realized how cruel it was I displayed so little interest in their lives before Mars. And they could probably see my little interest wasn't genuine, only polite. So after hearing about the house I asked for more.

"Tell me about the requirements for Mars One."

Why had I never asked that either? They gladly went through the whole, long, demanding procedure in detail, me smiling and happy to hear it--genuinely happy to hear it--and interested in every detail.

Preliminary Mars One qualifications: 18 years-old. A2 English. Resilient, trustworthy, adaptable and every other honorable human quality you can imagine. Then the training, so they could stay alive on their own so far from home: In-situ Resource Utilization: Atmosphere Management (gotta breathe), Wet Waste Processing (gotta waste), Thermal Control (mammals gotta stay warm). These are complex systems that have kept them and us alive all this time because our parents learned to create them and maintain them without failure for two decades.

"We had to know how to do it all," Dad said.

I wanted to grab them and hug them both and never let them go. Instead, I kept my voice from cracking and said, "I never realized how interesting...tell me more about where you and Chris Chang hung out, Dad."

That made him happy. Okay, sure, I'd hear about Sal's mortadella again but I'd also hear other things he'd never told me probably because I didn't look all that interested. Good Dad, kind Dad. Grinning happy Mom. The joy in telling me about Earth life, Jersey life, was evident in both of them and I hid my shame for not asking long ago.

I did not cry.

This went on through the whole evening and when eventually they were "getting sleepy" I said, "Sit with me for a selfie."

Everything is monitored. I could replay that whole evening anytime I wanted, but I wanted my own shot of us together taken at my own arm's length.

After that, we hugged and kissed good-night and I walked away from them toward the airlock door, tears spilling down my face, having a hard time breathing. They couldn't see that.

61

Next day a preliminary meeting for the whole colony was called in the infirmary to see if they remembered what they signed off on twenty years ago. The new batch that arrived ten years ago would remember, but they might not know it applied yet. I was steeled against what I was going to hear. It was certainly going to be delicate news to deliver.

Chang said, "Please, everyone, attention."

Patients were taking everything lightly that day including their crowded infirmary.

"Quiet, please, everybody," he said, but parents lounging and sitting on their beds went on talking. It was a party atmosphere.

Lewis stepped up and said, "Be quiet, everybody!"

They didn't often hear Lewis's voice and it got their attention.

Rob and I were watching from the side. "Glad this isn't me," I said.

Chang started. "It's good to see all of you together. Welcome. Just about everyone in the colony here..."

Rob whispered, "How's he going to do this?"

Chang took a deep breath, "It is a very serious matter that brings us together..."

Some wiseacre laughed and said, "Better be, getting us crammed in here like this."

A big laugh.

Chang smiled and said, "Well, it is," with a nervous laugh of his own. "The quake shook us up quite a bit and um...our biosphere is not in working order any longer."

That got no response. Chang went on.

"We won't have enough food to..." He didn't finish.

There were ohs and ahs but no despair.

"He hasn't figured out how to do this," Rob mumbled.

The parents look perplexed, not worried, and showed no indication they remembered the NASA imperative.

"They don't realize we're in danger," I muttered.

The others, the second batch of settlers who were not so afflicted, were aware and looked uneasy.

Chang said, "Running out of food isn't a pleasant end and we have an alternative."

"How 'bout no end at all?" said the same joker.

"Whose dad is that?" I said.

The parents shushed him and Chang answered.

"That's not possible. We will...we're all going to..."

Rob muttered, "Say it, Chris."

Parents frowned, perplexed, listening closely to what must have sounded confusing to them.

"It will be entirely up to you, each of you how, you deal with it. You can talk it over with your friends...or not. Your decision will be secret...or not. However you want it. It's up to you. Each of you will decide..."

"They don't know what he's talking about," Rob said.

The parents weren't disturbed. They gradually started talking, solemn but apparently not bothered by news they couldn't grasp. However, those less severely afflicted began drifting over to Rob and me.

"What is this? What does he mean?"

"Just what he said. The biosphere's function is breached. We can't keep growing food."

One of them said it. "We'll die?"

"Well...yeah."

Questions came at us all at once.

"When did this happen?"

"Why didn't we know?"

"Isn't the work rotation just a decision by the council?"

To this small group we explained in detail how the settlers went from irritability and conflict to eventual degeneration and now childish behavior.

"You watched it," I said. "You saw how they were and how they are now so...we'll all eventually be useless, no work will get done and no food will grow. We can hang on for a while but some of us need to survive. You do know that, right? You remember?"

They frowned and nodded. Some talked among themselves.

Behind them, somebody among the still heedless parents called out, "NASA can send food."

Chang said, "There's no supply probe that can carry enough for all of us and it would take months to get here."

"Isn't he going to tell them?" Rob said.

It didn't look like it, then Lewis stepped forward and I said, "Okay, here we go."

"We won't survive," Lewis said. "The entire colony is threatened."

There was more talking and moving around. Lewis raised his voice.

"Some of us have to survive, remember? And the only way to guarantee that happening..."

Parent voices were raised and they heard nothing more. Neither did we, drowned out by childlike parents talking loud, paying no attention to him. The meeting was falling apart. They seemed to have forgotten what Chang said or didn't care, or couldn't.

Sylva was among them trying to calm them when one called

out, "Trampoline!" and that did it.

The meeting broke up and a noisy, enthusiastic bustle filled the room.

Sylva stepped aside and let it happen. All the afflicted parents suited up and walked out, Dolf with them to monitor. The others with us who were appropriately disturbed, went back to their habs and the meeting dissolved.

Rob, Lewis, the doctors and I stood in the emptied infirmary, looking at each other.

Chang said, "I'm sorry, I..."

Lewis said. "NASA should have included a protocol for announcing it."

Chang raised his eyebrows and nodded.

62

Jeff at JPL was updated.

> MARS: You should send food asap. We know what will have to be done. As you know, some have already done it, step outside not wearing protection. We're still rotating the work stations. Don't want anybody else affected any sooner than necessary. Chang says he's measured the rate of the effect and it's increasing, damaging us quicker. What took years for our settlers will take months for the rest of us even with reduced hours in the bio.

"Jeff will see what we're doing," Lewis said to Rob and me. "Holding on long as possible. The final order will come from NASA when there are no more options."

JPL: Your surrender to the problem is not acceptable, Lewis. We have the best minds working on it. I don't mean only the best NASA minds. That goes without saying. We've got think tanks, inventors, scientists, writers, any group or person world-wide that might come up with an idea working on it. The website we put up crashed in an hour. More servers. It crashed again. It's up and running now. You're not first on national news with our weather troubles but you're always second. Both very important to us. Not sure how people will take your giving up. We've got our problems here but, Dude, it's shameful to surrender. Don't do it. There's a whole sector of the population who believe in you guys and Mars and what you've done for 20 years. You're giving us courage. You've (we've) got to find a cure. Take a look at the inbox on the Mars 20 website. Tons of bottom-up ideas from people all over the world.

MARS: Thanks for the good words but you read the inbox and let us know if any ideas are worth a try. I can only say that you're not talking to astronauts when you talk to us. They volunteer to be reckless. We didn't and don't. Forward workable plans only, please. Also, enough servers, Jeff? Really?

***

Rob and I sat watching Earth news, water rising and inundating everything for miles. There was no gradual border to the flooding. Only an endless view of the tops of houses, floating and sunken automobiles, the tops of trees to the horizon.

The announcer stood at an interactive map with a pulsing red line marking submerged areas.

"The entire Gulf coastline is under water and has been for over two months," she said. "New Orleans is still flooded, ever since Linda and Marguerite hit in rapid succession and ruptured the new flood walls. Those residents are being permanently resettled in the Midwest and western states." She tapped the map and it went to close-up.

"Flooding is so severe in Mobile, Pensacola and Panama City that FEMA is moving families out permanently. They are being resettled in cities farther inland."

The president's face came on the screen, a short clip.

"...this flooding is no longer seasonal and these areas are no longer safe, habitable locations..."

***

At JPL Jeff sat reading a message from me.

MARS: In the classroom this morning your flooding news was on the wall and the children watched. When it ended Lewis told the kids to come up with ideas to help the Earth victims. In half a morning there were plenty to choose from. We are forwarding this one. It is the idea of Gaby, eight Earth years old, and her twin Noelle. They studied a map of the world and dictated to their older sister Amelie what the U.S. could do to stay high and dry. From Gaby and Noelle to the people of Earth, especially those being flooded, a suggestion: "You should go to the Atacama Desert. No one lives there. It is just like Mars but it has air to breathe and we don't, so you wouldn't have to wear face glass. The weather there is very nice too. Most of the time it is warm so you could wear Earth clothes. There is not much water but you can use water reclaimers like we do. And you can live in habs like ours. You will like them. If ever we go to Earth we would like to stay in a hab on the Atacama. It will feel like home. It is very beautiful in the pictures. We think you should move away from the water to the Atacama."

They're just sayin' Jeff.

63

I went running over to the lab with my newly discovered notes, excited.

"Look at this," I said.

Lewis was filling half a wall screen with equations.

I insisted, "Look."

"What is it?"

"A source of cement."

"Not possible." He kept working.

"Look." I shoved my pad under his eyes.

"What is it?"

"Burn limestone, get cement. See that?" He glanced at the paragraph under his nose. "What does Mars have, Lewis? Lots of rock. Carbonate rocks. They're like clastic sediments, sandstone, sand grains, a mud clay matrix. Read it. See what it says? A crystalline cement is produced during diagenesis, right there."

He looked at it and frowned.

I kept going. "Okay, I'm not a geologist, but see what it says? Our diagenesis is fire, so burn our rocks and they'll make cement."

Lewis didn't move, just stood thinking, looking at the pad's formulas and diagrams.

"It's a good idea," he said. "A really good idea. But it won't work here."

"Why?"

"We'd need heavy machinery to gather enough rocks, and huge indoor firing furnaces to create even a small amount of cement."

"We only need a small amount to cover the cement that's

there."

"No, no. That cement has to go, all of it. And it has to be replaced."

"Like...build a new bio?"

"Like build a new cement understructure. Big job. Thousands of kilos of supplies from Earth in probes like they sent for Mars One. It took years to presupply that first voyage. They could send it again but it'd take months to get here and we'd still have to build it. That small amount won't help us."

"That's it? It won't work?"

He shook his head. "Really good idea, though."

As I started walking away he said, "The Atacama idea is great."

"It wasn't mine."

***

Another meeting of the entire colony was called for late afternoon, this time in the council, plenty of room there. The parents trooped over in a good mood, a third of our colony chatting and laughing. I'd never seen them so happy. The rest of us were somber.

"What are they doing to feel so good?" I asked.

"Milk and cookies," Chang said. "And lots of trampoline."

"They're going to die. Thomas is going to announce the

number of days we've got left and remind them of what they signed."

"I heard your dad say that what happens, happens, Kit, although maybe he wasn't talking about the bio and food."

"The trampoline, maybe?"

"Oh, no. If something bad happened to the trampoline they'd be ranting."

Rob was listening and said, "The Atacama is looking good, isn't it?"

We took our seats.

After a few standard colony updates, the new list of work assignments was handed out: No one working in the bio, only a few who were assigned to salvage and freeze the remaining edible foods. The rest of us were shifted. A few from engineering were moved to council. So were a couple of nurses including Dolf. He wouldn't be needed at the door any longer. Rob and I were moved from bio to council which was handy for communication with JPL and NASA. Lewis and Derek stayed in the lab, too important to move.

"We have approximately seven weeks here until we go on harsh rations," said Thomas, reading off his notes, well prepared. "After that, we have only a couple of months."

The parents murmured back and forth, still oblivious.

Lewis stepped up beside Thomas, no notes, the firmer voice.

"After those seven weeks the bio won't function," Lewis said.

"The plants will die when we can't tend them and eventually we'll all be at risk. Do you understand? NASA can't send enough food on a regular basis." They were quiet, taking it in. "We have seven more week then rations. And then..."

Somebody broke in, "Why did you let this happen?"

Somebody else said, "Can't you think of something?"

I think they understood very well, maybe even remembered what they'd signed, because a strong dad's voice called out, "We'll go back to work there. That'll take care of it."

They immediately started filing out of the rows of seats toward the door, presumably toward the biosphere to solve the problem.

Chang called after them, "It won't help."

Dolf and others blocked the door.

"Please," Chang said. "Come back and sit down. There's nothing you can do right now."

But they weren't hearing him. They assembled at the blocked door, restless and confused but determined to follow whoever it was that said they'd work in the bio again and everything would be all right.

"Let us out!" That strong dad's voice again. "You can't keep us here."

But Dolf barred the door.

They raised their voices, didn't like being penned in.

"Let us go."

"We want to go."

Lewis took over and said, "Sit down, all of you. There's no reason to leave. There's nothing you can do about this now..."

But they were scared which manifested as anger and it started to look like another brawl might break out. Nobody was throwing punches yet but they were bumping against each other and yelling to be released.

Doors were locked, nobody allowed to leave.

Chang was saying comforting words but they weren't listening and it became more unruly. They began weeping and wailing, not fighting, but feeding off each other's anger as though it made them feel better.

I saw my parents, both involved, Dad shouting something, then embracing Mom who was crying.

Mr. Elkins stood with Lewis, Derek, Rob and me and those who were not afflicted and the others not involved in the revolt. We were a small number.

Elkins said, "Glad I can't see this. If we need to run will somebody guide me?"

"I'm here," I said, and put my hand on his arm.

64

In the uproar, we almost missed it--a quiet, almost imperceptible reverberation.

None of the parents noticed. Their noisy scuffling masked it. But Lewis and I noticed and so did several others. It was slight compared to the ruckus going on around us with the rebelling parents beginning to move as a mass, indignant and looking for escape. Then another jiggle, bigger and definitely noticeable, stopped them.

"Aftershock," Lewis said, loud.

The parents looked around a few seconds, then went back to their row.

Rob said to Lewis, "When do aftershocks end?"

"Hard to tell."

Next came a rolling that was like before the quake.

Lewis looked around. "I don't like this."

And then a tremor.

"Another quake?" I said.

Lewis didn't answer. He went to the back door. I handed Mr. Elkins over to Dolf, and Rob and I followed Lewis out.

"Will JPL be any help?" Rob said as we jogged toward the lab.

"In fifteen minutes maybe."

There was nothing unusual to see. The shuddering had become very light.

I said, "I'll take a look."

"Where you going?"

I pointed up.

Jetpacking toward the pole my view was clear but I wondered if the camera in my face glass could pick it up enough for Lewis to see.

I went down lower, circled, and said to him, "You seeing this?"

"Yes, I am."

The distant underground thermal springs were geysering water, not black sand or dust. Water. And it wasn't coming from the pole or Olympus. It was coming from release points, deep cracks maybe, only 100 kilometers from the colony. Close. Scary.

"That's water, Lewis. Do you see it?"

"Yes, I do."

No time for cheering. This was a new problem. How to wrangle the broad, advancing flow of water into the colony or around it. Springs don't stop flowing and sometimes they get stronger. We had no model for what flowing water would do on Mars.

"It could flood us," Rob said.

We weren't ready, didn't realize how it would behave or how big a flow it would be. Springs were spouting higher than the barchan geysers. They could reach me and I was flying at 75 meters. Our sats would pick them up.

I said, "What do I do, Lewis? Tell me what to do."

"Follow the flow," he said. "I need the velocity and location. Stay over the edge."

I did just that and he read off my data and ran the numbers. For quite a while there wasn't another word from either Rob or Lewis.

I said, "We did good, Rob."

He gave me a quick "yeah" while Lewis never moved his eyes off the readouts from my telemetry.

"It's moving straight toward the colony, you can see that," I said.

Lewis said, "After-shock must have kicked it loose."

"It's not freezing, see that?" I said. "It's steaming and moving."

"Source is boiling...looks like it's got hot rock fragments keeping it liquid..."

It was flowing partly in the dry waterbeds but most of it was flooding across the ground picking up rocks and asteroid debris creating a rope of advancing Mars crud with hot rocks in hot spring water. This wasn't cookie dough. It looked like the foam at the edge of the surf on the virtual Riviera beach only bigger and thicker and unlike the surf, it was not going to recede.

"Surface debris is slowing it," Lewis said. "It's less than 20 kph now, you're 200 kilometers away."

"We've got 10 hours?"

"More. It'll keep picking up debris and slowing."

"Will it stop?"

"No way. Springs keep flowing, keep pushing. They'll flood over any debris lines that stop. I think we have a flood problem. Yeah, definitely a flood problem."

***

Fifteen minutes later at JPL in California, Jeff was groggy and half asleep saying, "Please tell me you woke me for something important."

It was very early morning and hoping to get a few hours sleep, he had an eager physics student earning extra credit on watch while he managed a nap.

"This is pretty unusual," the student said as Jeff shuffled in. "You see there where..."

Jeff leaned close to the screen and his blurred eyes widened. He blinked. They focused and his jaw dropped.

The student went on, "You see that line of..."

"I see it."

Jeff stared as he unconsciously eased into the seat at his console. My delayed jetpack camera view was on his screen. He saw gushing water.

"Fifteen minutes ago," Jeff muttered under his breath.

"Can't speed up light, sir," said the student.

Jeff didn't answer. He stared.

***

In Houston the NASA Mars unit were glued to their screens, questioning Jeff in Pasadena.

"The asteroid caused it?"

"Must have."

"How fast is it moving? Did they say?"

"Twenty. It's 200 kilometers away."

"Colony might have a flooding problem."

"Definitely."

"What can we do?"

Jeff said, "Take pictures."

"Seriously."

"I am serious. I'm betting they come up with something and it might help your gulf coast people."

"Larger flood there, Jeff. And it'll recede."

65

I jetted back to the colony to help.

"We need to direct the flow," Lewis said.

"Into the dry river beds?"

"Of course."

Rob said, "What if it gets stronger? There may be more

springs rattled loose."

Lewis said, "Probably are, or will be, but we'll be okay if we channel it like rivers. Those dry riverbeds are deep. They've handled water before. We'll dig as many channels as we have to."

"We've got to get it into the bio, remember," I said.

"We've gotta get it under control first."

"Where's the manpower for that?"

There was no answer and I sprinted to the infirmary.

***

"Get into your skins," I said.

The parents were lounging, talking about the disturbing running-out-of-food news. They didn't move.

"Quake's over but we need your help...put on your skins. Right now! Do it!"

I was their hero, shouting at them, and they listened. They started moving so I went calm and firm like Lewis. I flashed the video on their walls.

"Y'see this?"

The water was moving toward us, working its way fast over rocks, thickening with debris.

"Like surf," a parent said softly.

"No, it isn't," I said. "It's coming from springs not that far away. It's the water we need but it'll flood us unless we direct it.

Do you understand that? It's a cure to the problem at the bio but we've go to control it."

"Yeah, duh," said the comedian dad. (I never found out who that was.)

They talked nervously among themselves.

"What do we do?" my mother asked, calm and businesslike while Dad shushed the talkers.

(Thank you, Mom. Thank you, Dad.)

"We go up there and dig channels," I said. "We direct that water flow into the dry river beds, the ones beside the colony. If you can't do it, stay inside. Our habs are strong and water tight. But if you can go out there and help, we can control the water and capture enough to fix the bio."

"Then we'll be okay?" Mr. Elkins said.

"We'll be okay." My voice and attitude were strong, don't know where that was coming from. "Do you understand?"

I waited but nobody moved. I wondered if they didn't get it, it was just too much for them. Then I heard some of the men begin talking together, very intense.

"You should be listening," I said. "This is life or death."

They paid no attention, kept on talking, nodding, short exchanges back and forth.

"You should be listening, this is urgent!"

"We know that. You're saying the water can save our lives?"

"That's right."

"You need workers to control it?"

"Right."

They looked at one another. "Well, all right then. Let's roll."

They pulled on skins and started marching out. The other men followed and so did the women right behind them, motivated again by their sense of solidarity and group can-do.

***

Rob and Lewis saw this on screens inside the lab and came out to watch. I came from the infirmary and joined them.

Lewis said, "What just happened? What are they doing?"

"They think they're going to direct the flow. They have some kind of plan, I think. How long before it has to be channeled?"

"Several hours," Lewis said. "They'll have to know what to do and where..."

He hurried back into the lab to do the calculating.

Meanwhile, the parents and some of the older teens all equipped with picks and shovels appeared in their rovers driving together behind a self-appointed leader: a parent with a torn white tee shirt stretched over the top of a curved pick-ax held upright outside the rover. The faded blue letters, YANKEES, spread across the shirt and led the convoy out past the dry river beds toward the still unseen incoming water.

"Tell us where to go," one of them said to Lewis, face-glass connecting us all.

Lewis said to Rob and me, "Go up and see where it is and get me the speed."

We jetted to the moving water line and provided the data. Knowing the speed, Lewis could tell the parents who were well on their way the best place to start the intercept and dig channels.

"Make it a ditch, deep if you can," Lewis told them.

All of us, Lewis in the lab, Rob and me in the air, and every person in a skin and face glass loaded with cameras were talking back and forth, seeing what we needed to see and hearing what we needed to do.

I asked Lewis if the water would "carve a little creek and then a brook and then a stream like on Earth? Or will it overflow?"

"It should follow like a stream but I don't know," he said. "That's what we're going to try to make it do in Mars-g. Keep watching and keep us on course."

My jetpack telemetry continued to inform Lewis and he informed the diggers.

"I'm going back to help," Rob said and flew away.

66

Farah Noor, ex-Los Angeles civil engineer, took charge of the team of diggers, automatically and naturally. Working with her was Tom Lester of St. Joseph, Missouri, an ex-city construction engineer, both skilled at crisis management.

"We know what to do here," Farah said and the parents listened.

Following instructions and preceding the diggers, a team of strong men with picks would crack through the permafrost for the parents who would dig trails of ditches for the ever-creeping-closer flow. Farah and Tom huddled with calculations provided by Lewis and plotted the design. These ditches had to hold water but they didn't need to be deep they said. They instructed the diggers who listened, asked a few questions, then went to work.

Farah and Tom went back to designing a further course leading to the riverbed near the colony. They stayed in touch with Lewis who fed them ongoing data from my aerial view so they could adjust their plan. The diggers, our parents, simply went on working like automatons, tireless robots.

It was clear this was going to take hours, all night. So runners, younger kids in their rovers, proud of the responsibility, brought water and sandwiches to the parents who'd been digging with little rest, both men and women. The diggers got in the rovers (they're little airlocks when needed), pushed back their hoodie-helmets, and ate and drank. When finished, they locked down their helmets again and went back to work. Eventually, at some point, each of them stepped aside for a rest, but they were always soon back at work. There was no keeping them down.

They heaped their shovel loads carefully to the side of the ditch as per instructions, building a small wall. No doggy digging like in memory pieces, throwing dirt out in a spray. They had a job. They were determined to do it right. This was the first time there was a need for their decades of strength maintenance. And it was the first time we noticed it.

I couldn't imagine how they'd keep going at this rate, but they said, "Don't worry...we'll do it."

Their muscles and bones and lungs are more developed than ours after a youth in Earth-g and years of workouts in Earth-g weights on Mars. Also, their short recovery time was something we'd never seen. Amazing.

Their leader, unstoppable Jakob (Moscow Institute of Physics and Technology, "Go Bears"), was their coach and he set the standard.

"Get rhythm," he called out. "Keep rhythm."

He set the cadence for shoveling with a rocking meter followed by short rests then more rhythmic digging. They worked this way through the night, men and women, bodies bobbing and rocking, lights from their face glass flashing.

Following instructions, they created a pattern of straight ditches leading to angled ditches leading to more straight and angled ditches making more spidery brain circuitry across the regolith that led toward the colony and our dry riverbeds. Hopefully when the water reached these ditches, and it was getting close, it would flow into them and stay in them all the way.

Occasional views that Lewis sent to everybody's head-up display showed the water's edge creeping closer. After several hours, most of the night actually, the rope of water and junk so like a surf's edge was very close to our beautiful intricate pattern of channels. Once the water was caught up in those, we could only hope it would flow as planned. We had to be ready if it didn't.

Tom and Farah consulted with Lewis.

"Two things," Farah asked. "Where do we lead the water once we've got it near the colony? And what do we do if the water overflows the riverbeds and our ditches? It'll flood the colony and getting it into the bio will be impossible."

For a second I tried to picture the scene at NASA 15 minutes behind everything we did, everyone crowding the room gaping, minds boggled hearing our conversations--flowing water on Mars for god's sake, international news, and yet possible disaster for the colony.

Directing the water was only the beginning, not the full solution. We could still fail like in all stories. Little Red Riding Hood was fine until the wolf showed her a better path to Granny's house, then it all went terribly wrong.

"We need a plan," Lewis said to me, to everybody listening wherever they were.

"Think," he said.

Think is a compelling order. In this case, it was for life or death and a long silence followed.

I had an idea.

***

A second unit of diggers was sent to start another off-direct.

"If there's an overflow near the colony we can dig a canal right through the center of the colony and out the other side," I said. "No flooding, no damage."

Tom and Farah said it would work but required more workers than we had. The ones already digging couldn't give up any hands. I went down to help, and everybody else came out and pitched in. Lewis and Mr. Elkins stayed in the lab with very small children who were curled in blankets on the floor sleeping peacefully. We had a small, fresh workforce.

"We need some kind of half pipe," Farah said.

We have plenty of pipe--two eight hundred meter shafts of cylinder, plus our standard supply of polymer amalgam pipe for use in the habs. Lewis cut the cylinder segments lengthwise and then into short lengths and Rob rovered a supply out to the new diggers who were gathered, waiting.

We laid the pipe starting at the ideal entry into the colony then back up toward the ditches already dug. They connected to river beds where our channeled water would empty and soon be flowing, we hoped. That went fast. Soon as it was done our second unit of workers went on up to help the flagging, all night diggers who still insisted they weren't flagging and didn't need help.

"We're okay," they said.

"You've done it," we told them. "We've got pipe waiting below."

They were dopey with fatigue but pleased at their accomplishment though a few simply folded to the ground. We took over for them and they rovered back to the infirmary. Rob and I helped with the last digging but we had to rest twice as often as the parents and for twice as long.

"They're amazing," I said and wondered how they were doing.

I contacted Sylva.

"They're fine," she said. "They're watching MacGyver."

As Rob and I rested I said, "If we live through this, do you think we should take a look at the MacGyver thing?"

67

The worst was over. We'd done it. The work that followed in the morning kept the water from inundating the colony. It overflowed here and there and we dug small ditches to correct the problem while the rest of the water followed the well-designed ditches and entered near the bio. Or would enter when we wanted it to. For the time being, it was shunted into the riverbed.

It flowed there for the next few days while we created a regulated flow through the biosphere basins. That meant modifications of the biosphere itself. (Lewis will forward technical details on that construction.) Here, in brief, is what we did.

Our most plumbing-versed settlers, Boa Wing and Gunter Bauer (all these experts deserve credit) went to work installing pipes that would bring in the water, direct it through the basins and out the other side of the dome.

That required connecting all the basins into the main continuous flow line. When that was done and the new system was in working order, the problem was solved.

The biosphere was repaired, safe to work in and no one would be harmed. We would not die.

***

The parents' health and general condition were greatly improved by their efforts. They were more alert, eager to start work early each morning because they had an important job to do. Lewis worked out algorithms to take advantage of the temperature and varying force of the flow of Mars water into our bio.

Several parents took on the job of water channel maintenance. They decided to rotate it with others and to have a couple of monitors to make sure it's done correctly. I feel certain that we have, without a doubt, the finest water channel maintenance crew on the planet, possibly in the solar system.

But that important job doesn't require all the parents and some are freed up to go back to work in the biosphere. It's the place where they originally chose to work and liked working until things went wrong.

So our workforce is complete again, everybody content. The few who chose to go back into the infirmary are older and have decided to live out their natural lives there watching videos of their amazing work on the ditches and their creation of the pipeline to the colony along with their own memory pieces and, for the occasional break, MacGyver.

They watch the pipeline digging over and over and marvel at what they did, then they congratulate themselves and sleep soundly.

With everything in place it was decided to have a celebration.

***

The colony's common area was crisscrossed with crepe paper streamers. We make those here. Our crepe paper did nothing unusual in the sub-zero temp except shatter if it fell. No particular use for that information on Earth, but as instructed, we pass on all our findings.

Celebration day was grand. Lewis gave a nice talk about the

good work everybody did, and how much is owed to the parents.

"You literally saved our lives," he said. "I tried to make a list of names to thank, and realized it's all of you, everyone, even our young shuttle rover drivers."

Those kids cheered for themselves.

"We're definitely better for what we went through," Lewis said. "I see faces standing out there that were bedridden before all this..."

The much-recovered parents gave themselves applause and the rest of us joined in. Farah and Tom were lifted onto shoulders as was the convoy leader with his YANKEES tee shirt, and master digger Jakob, of course, with his tee shirt worn over his skin gear, a bear on the front.

Lewis said it again, "You saved our lives," and we cheered.

"So did you, Lewis," Rob shouted, and everybody cheered him.

Then somebody tried to get us singing For He's A Jolly Good Fellow but Gen2s don't know the words all that well and it dribbled into laughter and applause.

Every group--rover kids, experts, diggers, plumbers--was happy. Settlers are our heroes. Without their bodies in their physical condition, we couldn't have done it.

At one point Lewis said to the parents, "I've made your kids promise to never question you or look bewildered again when you talk about Earth."

That got a big laugh and a bigger cheer. Then seeing the jubilation wasn't dying down, Lewis put our favorite music blaring from every screen and speaker in the colony (the right way to use audio, Mr. Thomas). We danced and celebrated in front of the biosphere and, did I mention? Baby Bots 2 and 3 were there. Lewis wanted to see what they'd learn.

They responded to the music. That got them going, both keeping time to the beat. They must have "felt" it, registered it, seeing us bobbing around. So they were nodding and bobbing with us in perfect rhythm, occasionally saying "Good morning" to anybody nearby.

Lewis was grabbed into the dancing crowd and forced to join in which I did not like at first thinking it made him uncomfortable. But as I watched, concerned for his state of mind with parents nudging him and saying, "Go on, go on," Lewis dropped and twisted and did some kind of fantastic acrobatic dancing and spinning and back stepping that I can't describe. For the purpose of this report let me say he blew our minds. Everybody backed away to watch, then we yelped and applauded and made him do more.

For once, I wanted to be on Earth watching that. He's no longer just our unparalleled physics guy even though he's gone back to that persona. We know now there's somebody else underneath and we like that guy too.

Because the day was memorable, Gen2s now have our own

memory piece, the record of all that happened that celebration day taken from angles all over the common area. We'll be adding to it with everything that happens from now on. We're making our own history, to be added to that of our parents, of course.

This ends the official calamity report.

<u>One last thing</u>. Do not delete this.

We have been informed that since the asteroid impact, there is excitement on Earth about the value of the raw materials in the asteroid buried on Olympus Mons. We object to your sending those who want to exploit those riches, as you call them. We've been told we would be expected to create some kind of banking system to handle the percentage of profits all Martians would receive. But we have no need for profits or banks. We called a quick assembly at the council and every single colony member of age agreed. We vetoed any mining of the asteroid other than Martian. It brought us flowing water and is a natural mineral resource for us which lessens the need for so many supply probes from Earth. That is an advantage to all.

We suggest you visit www.Expedia.Com/daily/mars/mars-activities/ where there is a full-color section on Mars that includes sports and day trips to our many sites, the Galle Crater, Utopia Planitia, and the very impressive Valles Marineris--you think *your* Grand Canyon is big?

Also as soon as there is a shuttle service to Earth, we

recommend our planet for your holidays or even as a permanent home. We are the perfect retirement community with beautiful views and crisp, occasionally dusty weather. You will find yourself accepted in our colony and we look forward to your visit. In fact, on your long, final approach from space, make sure you watch from a viewing port in your spacecraft. At first you will see it as a white spot from as far out as two light minutes, but keep watching and as you near, it will become clearer and clearer what the dot is saying.

WELCOME TO MARS

PS: The twins urge you to seriously consider the Atacama for your current flooding problems.